Arthur S. Edding

The Theory of Relativity
and its Influence
on Scientific Thought

Selected Works on the
Implications of Relativity

Edited by Vesselin Petkov

MINKOWSKI
Institute Press

Arthur S. Eddington
28 December 1882 – 22 November 1944

Cover: The title page of the original publication of the Romanes Lecture 1922: A. S. Eddington, *The Theory of Relativity and its Influence on Scientific Thought* (Oxford University Press, Oxford 1922).

ISBN: 978-1-927763-32-2 (softcover)
ISBN: 978-1-927763-33-9 (ebook)

Minkowski Institute Press
Montreal, Quebec, Canada
http://minkowskiinstitute.org/mip/

For information on all Minkowski Institute Press publications visit our website at http://minkowskiinstitute.org/mip/books/

Arthur Stanley Eddington in 1931
(Source: http://silas.psfc.mit.edu/eddington/)

Has not a deeper meditation taught certain of every climate and age, that the WHERE and the WHEN so mysteriously inseparable from all our thoughts, are but superficial terrestrial adhesions to thought?

CARLYLE, *Sartor Resartus.*

Preface

This volume contains new publications of six works by Arthur S. Eddington – five directly exploring the implications of the theory of relativity and one (The Decline of Determinism) indirectly related to it – and a Symposium on the philosophical aspects of the theory of relativity with the participation of two physicists (A. S. Eddington and F. A. Lindemann) and two philosophers (C. D. Broad and W. D. Ross):

1. A. S. Eddington, *The Theory of Relativity and its Influence on Scientific Thought* (Oxford University Press, Oxford 1922). This is the Romanes Lecture 1922 – Delivered in the Sheldonian Theatre on 24 May, 1922.

2. A. S. Eddington, The Relativity of Time, *Nature* **106**, 802-804 (17 February 1921)

3. A. S. Eddington, The Meaning of Matter and the Laws of Nature according to the Theory of Relativity, *Mind* **29** (114) 145-158 (1920)

4. A. S. Eddington, The End of the World: from the Standpoint of Mathematical Physics, *Nature* **127**, 447-453 (21 March 1931)

5. A. S. Eddington, The Expanding Universe, *Nature* **129**, 421-423 (19 March 1932)

6. A. S. Eddington, The Decline of Determinism, *Nature* **129**, 233-240 (13 February 1932)

7. A. S. Eddington, W. D. Ross, C. D. Broad and F. A. Lindemann, The Philosophical Aspect of the Theory of Relativity: A Symposium, *Mind* **29** (116), 415–445 (1920). Published as an Appendix.

Montreal, 5 January 2015 *Vesselin Petkov*

CONTENTS

Preface   i

Introduction   1

The Theory of Relativity and its Influence on
Scientific Thought   7

The Relativity of Time   27

The Meaning of Matter and the Laws of Nature
according to the Theory of Relativity   31

The End of the World: from the Standpoint
of Mathematical Physics   45

The Expanding Universe   59

The Decline of Determinism   65

APPENDIX: The Philosophical Aspect of the
Theory of Relativity: A Symposium   81

INTRODUCTION

Arthur Stanley Eddington was known not only as a world-renowned experts on Einstein's theory of relativity but also for his rare talent to convey the most profound ideas of modern physics to a wider audience. This volume reprints a collection of some of his best works mostly on the implications of relativity for our understanding of the world.

There are two main reasons for the publication of this volume. First, to offer to non-experts a collection of articles on the influence of relativity on our views of the world by Eddington – the physicists who in 1918 provided the first systematic exposition of Einstein's general relativity in English and who was involved in its first experimental test in 1919. Second, and perhaps more importantly, to provide young and future physicists with a collections of specifically chosen works by Eddington to see how one of the best physicists of the 20th century was viewing the advancement of fundamental physics in the beginning of the 20th century.

It is true that sometimes Eddington held controversial views. But I think even that side of Eddington might be beneficial for future scientists, because one can hardly have a chance of achieving ground-breaking results without contradicting existing views of mainstream science. Here is Eddington's own justification for this (p. 51):

> I have sometimes been taken to task for not sufficiently emphasising in my discussion of these problems that the results about entropy are a matter of probability, not of certainty. I said above that if we observe a system at two instants the instant corresponding to the greater entropy will be the later. Strictly speaking I ought to have said that for a smallish system the chances are, say, 10^{20} to 1, that it is the later. Some critics seem to have been shocked at my lax morality in making such a statement, when I was well aware of the 1 in 10^{20} chance of its being wrong. Let me make a confession. I have in the past 25 years written a good many papers and books, broadcasting a large number of statements about the physical world. I fear that for not many of these statements is the

risk of error so small as 1 in 10^{20}. Except in the domain of pure mathematics the trustworthiness of my conclusions is usually to be rated at nearer 10 to 1 than 10^{20} to 1; even that may be unduly boastful. I do not think it would be for the benefit of the world that no statement should be allowed to be made if there were a 1 in 10^{20} chance of its being untrue. Conversation would languish somewhat. The only persons entitled to open their mouths would presumably be the pure mathematicians.

Although the papers in this volume speak for themselves I will take the liberty of drawing your attention to several issues in Eddington's articles "The Relativity of Time" and "The Decline of Determinism" and to three points in the Symposium on the philosophical aspects of the theory of relativity with the participation of two physicists (A. S. Eddington and F. A. Lindemann) and two philosophers (C. D. Broad and W. D. Ross) given in the Appendix.

As the story goes, when Eddington was told (after his talk reporting on the first experimental test of Einstein's general relativity in 1919) that he was one of the three people in the world who understood Einstein's theory, he had reportedly wondered who the third person was. Quite possibly Eddington might have thought that that person would have been Hermann Minkowski (had he lived longer), because Eddington knew well that Minkowski had a profound understanding of Einstein's special relativity and certainly would have had such understanding of the general theory of relativity as well since that theory would be impossible without Minkowski's spacetime formulation of special relativity.

I think future physicists could particularly benefit from examining Eddington's view on the nature of spacetime in his article "The Relativity of Time" (1921) because now it has almost become fashionable (and perhaps convenient) to ignore this question by declaring it philosophical. Here is what Eddington thinks about the reality of Minkowski's four-dimensional world, i.e. about the reality of spacetime; he does regard these two notions as synonyms (p. 45): "the world – or space-time – is a four-dimensional continuum". Eddington first explained that special relativity taught us (thanks to Einstein and Minkowski[1]) that time and space are not absolute but relative (or fictitious), because (p. 28):

---

[1] It was Minkowski who first realized that if observers in relative motion have different times, they must have *different spaces* as well, which is *only possible in a four-dimensional world*: "Hereafter we would then have in the world no more *the* space, but an infinite number of spaces analogously as there is an infinite number of planes in three-dimensional space. Three-dimensional geometry becomes a chapter in four-dimensional physics." H. Minkowski, Space and Time, in H. Minkowski, *Space and Time: Minkowski's Papers on Relativity*, (Minkowski Institute Press, Montreal 2012).

The time determined by astronomers and in general use is thus a fictitious time, or, in the usual phrase, it is *relative* to terrestrial observers. Similarly it has been found that extension in space is also relative.

Then he directly addressed the question of the physical meaning of these spaces and times and especially the question of the reality of spacetime (p. 28):

It was shown by Minkowski that all these fictitious spaces and times can be united in a single continuum of four dimensions. The question is often raised whether this four-dimensional space-time is real, or merely a mathematical construction; perhaps it is sufficient to reply that it can at any rate not be less real than the fictitious space and time which it supplants.

I think it is obvious that Eddington dealt with the issue of the reality of spacetime not to please philosophers, but because that issue is an intrinsic physical one – it is about the *dimensionality* of the world and it is solely physics that should deal with it. Moreover, deeper understanding of the nature of spacetime (the four-dimensional world) might have implications for physics itself as Minkowski anticipated. After he had successfully decoded the profound message – that the world is four-dimensional – hidden in the failed experiments to detect absolute uniform motion he had certainly realized that four-dimensional physics was in fact spacetime geometry since all particles which *appear* to move in space are in reality a forever given web of the particles' worldlines in spacetime. Then Minkowski outlined a program for geometrization of physics in his talk "Space and Time": "The whole world presents itself as resolved into such worldlines, and I want to say in advance, that in my understanding the laws of physics can find their most complete expression as interrelations between these worldlines."

Had Minkowski lived longer he might have quite probably applied this program to gravitational phenomena by exploring the tempting possibility that gravity might not be a physical interaction since the apparent gravitational attraction can be fully explained as a mere manifestation of the non-Euclidean geometry of spacetime.[2] Not surprisingly Eddington might have also explored this captivatingly radical possibility – "gravitation as a separate agency becomes unnecessary" (p. 30).

Eddington's deep understanding of Minkowski's view of the four-dimensionality of the world should be taken into account when his

---

[2] See V. Petkov, Physics as Spacetime Geometry, in A. Ashtekar, V. Petkov (eds), *Springer Handbook of Spacetime* (Springer, Heidelberg 2014), pp. 141-163.

4

paper "The Decline of Determinism", which does not deal with relativity, is read. The reason is that it is often stated that spacetime cannot accommodate the probabilistic behaviour of quantum objects and, if that is correct, it would turn out that Eddington holds contradicting views. Eddington's paper "The Decline of Determinism" is included in this volume specifically to demonstrate that Eddington did not see anything contradicting to hold both views – that spacetime is real and quantum phenomena are probabilistic. I think his reason is exceedingly clear – both views are derived from the experimental evidence and *experiments do not contradict one another*. The view that probabilistic phenomena are impossible in spacetime is based on the implicit assumption that quantum objects are worldlines in spacetime. But the experimental evidence in quantum physics has ruled that out. And indeed quantum objects might be other spacetime structures, not necessarily worldlines.[3]

Finally, let me draw your attention to three points in the Symposium "The Philosophical Aspect of the Theory of Relativity" published as an Appendix with the participation of two physicists (A. S. Eddington and F. A. Lindemann) and two philosophers (C. D. Broad and W. D. Ross).

1. *W. D. Ross' difficulty with the notion of simultaneity.* It should be kept in mind that when Einstein demonstrated that simultaneity of distant events is relative, it became clear that the world is not objectively divided into slices of simultaneous events (as Minkowski powerfully demonstrated). This means that the expression "a class of simultaneous events" does not represent anything in the objective world and is nothing more than a human construction to represent the external world in a scheme which is implied by our senses. That is why it is now understood that Einstein had a deep intuition when he wrote in his 1905 paper that whether two distant events are simultaneous is a matter of definition (or convention). Indeed, if simultaneity is relative it is necessarily conventional and vice versa.

2. *C. D. Broad's confusion that Einstein's special relativity does not deal with accelerated motion* (p. 94):

> The special theory explicitly confined itself to systems in uniform translational motion with respect to a Newtonian frame of reference. It did not profess to tell us what would happen if a system rotated with respect to such a frame or moved with an accelerated rectilinear motion with respect to it.

It was again Minkowski who pointed out that acceleration is fully explained in the four-dimensional world (described by special rela-

---

[3]See V. Petkov, *Relativity and the Nature of Spacetime*, 2nd ed. (Springer, Heidelberg 2009), Chap. 10 and the references therein.

tivity) – he stressed in his talk "Space and Time" that "Especially the concept of *acceleration* acquires a sharply prominent character": a curved (deformed) worldline in spacetime represents an accelerating particle, whereas a straight worldline represent a particle moving with constant relative velocity (when the relative velocity of the particle is zero with respect to another particle, the worldlines of the two particles are parallel and the particles are at rest relative to each other).

The fact that the confusion that acceleration is not covered by special relativity was not fully eliminated prompted Misner, Thorne, and Wheeler in 1973 to devote a whole chapter on this issue in their *Gravitation*[4] – Chapter 6 (Accelerated Observers) whose first section is entitled "Accelerated observers can be analyzed using special relativity."

3. *F. A. Lindemann insistence that it is only a matter of convention whether we will describe gravitational phenomena in terms of general relativity or the Newtonian gravitational theory* (p. 107):

> The absolutist system introduces a mysterious entity called force and requires five assumptions at least. The relativist system yields all the same results with but three assumptions. The latter, therefore, appears preferable, but to say that one assumption is true and the other false would be just as meaningless as to say that space is or is not homaloidal. Either point of view is perfectly justified, but the one appears simpler, and, therefore, more convenient than the other.

Although it is now accepted that general relativity is the correct theory of gravitational phenomena one could still hear the phrase "physical theories are just descriptions" and it is up to us to decide which description to use. If such statements are made for specific cases – e.g., to describe a falling body (where Newton's theory should be obviously used) – that is understandable, but often it is meant more than that.

I think one should be very and constantly alert to statements such as "*it is just a matter of description*" if they imply that physical phenomena can be described *equally* by different theories. Such a view is dangerous since not only does it ultimately hamper our understanding of the world but also negatively affects the advancement of fundamental physics. I believe the history of physics has proven that part of the art of doing physics is to determine whether different theories are indeed simply different descriptions of the same physical phenomena (as is the case with the three representations

---

[4]C.W. Misner, K.S. Thorne, J.A. Wheeler, *Gravitation* (Freeman, San Francisco 1973).

of classical mechanics – Newtonian, Lagrangian, and Hamiltonian), or *only one* of the theories competing to describe and explain given physical phenomena is the correct one (as is the case with general relativity, which identifies gravity with the non-Euclidean geometry of spacetime, and other theories, which regard gravity as a force[5]).

Montreal, 5 January 2015   *Vesselin Petkov*

---

[5]The existing experimental evidence – that *falling bodies do not resist their accelerated motion while falling* – unequivocally demonstrates that gravity is not a force; if the falling bodies did resist their acceleration, only then a (gravitational) *force would be needed to overcome the resistance* (as Newton's second law demonstrates).

THE THEORY OF RELATIVITY AND ITS INFLUENCE ON SCIENTIFIC THOUGHT

A. S. Eddington, *The Theory of Relativity and its Influence on Scientific Thought* (Oxford University Press, Oxford 1922). This is the Romanes Lecture 1922 – Delivered in the Sheldonian Theatre on 24 May, 1922.

In the days before Copernicus the earth was, so it seemed, an immovable foundation on which the whole structure of the heavens was reared. Man, favourably situated at the hub of the universe, might well expect that to him the scheme of nature would unfold itself in its simplest aspect. But the behaviour of the heavenly bodies was not at all simple; and the planets literally looped the loop in fantastic curves called epicycles. The cosmogonist had to fill the skies with spheres revolving upon spheres to bear the planets in their appointed orbits; and wheels were added to wheels until the music of the spheres seemed wellnigh drowned in a discord of whirling machinery. Then came one of the great revolutions of scientific thought, which swept aside the Ptolemaic system of spheres and epicycles, and revealed the simple plan of the solar system which has endured to this day.

The revolution consisted in changing the view-point from which the phenomena were regarded. As presented to the earth the track of a planet is an elaborate epicycle; but Copernicus bade us transfer ourselves to the sun and look again. Instead of a path with loops and nodes, the orbit is now seen to be one of the most elementary curves – an ellipse. We have to realize that the little planet on which we stand is of no great account in the general scheme of nature; to unravel that scheme we must first disembarrass nature of the distortions arising from the local point of view from which we observe it. The sun, not the earth, is the real centre of the scheme of things – at least of those things in which astronomers at that time had interested themselves – and by transferring our view-point to the sun the simplicity of the planetary system becomes apparent. The need for a cumbrous machinery of spheres and wheels has disappeared.

Every one now admits that the Ptolemaic system, which regarded the earth as the centre of all things, belongs to the dark ages. But

7

to our dismay we have discovered that the same *geocentric* outlook still permeates modern physics through and through, unsuspected until recently. It has been left to Einstein to carry forward the revolution begun by Copernicus – to free our conception of nature from the terrestrial bias imported into it by the limitations of our earthbound experience. To achieve a more neutral point of view we have to imagine a visit to some other heavenly body. That is a theme which has attracted the popular novelist, and we often smile at his mistakes when sooner or later he forgets where he is supposed to be and endows his voyagers with some purely terrestrial appanage impossible on the star they are visiting. But scientific men, who have not the novelist's licence, have made the same blunder. When, following Copernicus, they station themselves on the sun, they do not realize that they must leave behind a certain purely terrestrial appanage, namely, *the frame of space and time* in which men on this earth are accustomed to locate the events that happen. It is true that the observer on the sun will still locate his experiences in a frame of space and time, if he uses the same faculties of perception and the same methods of scientific measurement as on the earth; but the solar frame of space and time is not precisely the same as the terrestrial frame, as we shall presently see.

I think you will readily understand what is meant by a *frame* of space and time. It is the system of location to which we appeal when we state, for example, that one event is 100 miles distant from and 10 hours later than another. The terms space and time have not only a vague descriptive reference to a boundless void and an ever-rolling stream, but denote an exact quantitative system of reckoning distances and time-intervals. Einstein's first great discovery was that there are many such systems of reckoning – many possible frames of space and time – exactly on all fours with one another. No one of these can be distinguished as more fundamental than the rest; no one frame rather than another can be identified as the scaffolding used in the construction of the world. And yet one of them does present itself to us as being the actual space and time of our experience; and we recoil from the other equivalent frames which seem to us artificial systems in which distance and duration are mixed up in an extraordinary way. What is the cause of this invidious selection? It is not determined by anything distinctive in the frame; it is determined by something distinctive in us – by the fact that our existence is bound to a particular planet and our motion is the motion of that planet. *Nature* offers an infinite choice of frames; *we* select the one in which we and our petty terrestrial concerns take the most distinguished position. Our mischievous geocentric outlook has cropped out again unsuspected, persuading us to insist on this terrestrial space-time frame which in the general scheme of nature is in no way superior to other frames.

The more closely we examine the processes by which events are assigned to their positions in space and time, the more clearly do we see that our local circumstances play a considerable part in it. We have no more right to expect that the space-time frame on the sun will be identical with our frame on the earth than to expect that the force of gravity will be the same there as here. If there were no experimental evidence in support of Einstein's theory, it would nevertheless have made a notable advance by exposing a fallacy underlying the older mode of thought – the fallacy of attributing unquestioningly a more than local significance to our terrestrial reckoning of space and time. But there is abundant experimental evidence for detecting and determining the difference between the frames of differently circumstanced observers. Much of the evidence is too technical to be discussed here, and I can only refer to the Michelson-Morley experiment. I fear that some of you must be getting rather tired of the Michelson-Morley experiment; but those who go to a performance of Hamlet have to put up with the Prince of Denmark.

This famous experiment is a simple test whether light travels at the same speed in two different directions. For this purpose an apparatus is constructed with two equal arms at right angles, providing two equal tracks for the light. A beam of light is divided into two parts so that one part travels along one arm and back, and the other along the other arm and back. The two rays then re-unite, and by delicate interference tests it is possible to tell if one has been delayed more than the other; a delay of less than a thousand-billionth of a second could be detected. The experiment is simply a race between two light-rays with equal tracks, but pointing in different directions; the result turns out to be a dead-heat. At first sight this is just what would be expected; and one almost wonders why it should have been thought worth while to try the experiment. But Michelson, like a good Copernican, had stationed himself on the sun to watch the race; accordingly he realized that the apparatus was being borne along by the earth's orbital motion with a speed of 20 miles a second. Consequently the light does not travel exactly the double length of the arm; starting at one end it has to go to the turning-mark at the other end which has moved on a little in the meantime; then it returns to the place which the starting-mark has travelled to whilst the race is in progress. That does not add up to exactly the double-length of the arm. Making the calculations we easily find that, although the two arms are equal, the two light-journeys are unequal; the competitor whose track lies in the line of the earth's motion has the longer journey, and is at a disadvantage. And yet according to the experiment he does not suffer the expected delay. From our standpoint on the sun, the experiment seems to have gone wrong; Copernicus has met with a rebuff, and Ptolemy is triumphant.

But that is because we have not admitted the full consequences of transferring our standpoint to the sun. We have all the while been keeping one foot on earth. Of course, the whole experiment turns on the two arms having been first adjusted to perfect equality. This could only be ascertained by experiment; and the test applied was to rotate the apparatus through a right angle, so that if, for example, the journey in the line of the earth's motion had had the advantage of the shorter arm on one occasion, the transverse journey would have had it on the repetition. That is a perfectly satisfactory test for a terrestrial observer; to turn a rod from one direction to another is the simple and direct way of marking out equal lengths. But the test is not satisfactory to an observer on the sun; he would not think of attempting to partition equal lengths of space by means of rods travelling at 20 miles a second. His frame of space – the space not only of refined measurement, but also of the cruder measurements made with the sense-organs of his body which determine his perception of space – is partitioned by appliances at rest relatively to him, e. g. his own eyes and limbs. Lengths of objects carried on the earth must be judged by him according to the room they occupy in his own frame. In the space of the terrestrial observer the two arms of the apparatus were adjusted to equal length; but in the re-partitioned space of the solar observer they may quite well occupy unequal lengths, and when we take the view-point of an observer on the sun we must not overlook this inequality. This inequality is not so much an hypothesis proposed to account for Michelson's result as a direct deduction from it. The two light-journeys were found to occupy equal times; this clearly shows that the arm in the less favoured direction is shorter than the other so as to counterbalance the handicap to which I have referred.[1]

When the apparatus is turned through a right angle, the experiment still gives the same result. It does not matter which of the two arms we place in the line of the earth's motion; that arm must be shorter than the other. In other words each arm must automatically contract when it is turned from the transverse to the longitudinal position with respect to its line of motion. This is the famous FitzGerald contraction of a moving rod. It is of the same amount whatever the material of the rod, and depends only on the speed of its motion. For the earth's orbital motion the contraction amounts to one part in 200 million; in fact the earth's diameter in the direction of its motion is always shortened by $2\frac{1}{2}$ inches, the transverse diameter being unaffected.

This contraction of a moving material object was first revealed

---

[1] The only alternative is that (relatively to a solar observer) the velocity of light differs in different directions, at least in the region where the experiment is conducted. This would presumably be due to some influence of the moving earth on the propagation of light (convection of the ether). This explanation was at one time favoured, but it could not be reconciled with the observed phenomena of the aberration of light.

to us by the Michelson-Morley experiment; but it is not at all disagreeable to theoretical anticipations. We have to remember that a rod consists of a large number of molecules kept in position by their mutual forces. The chief force is the force of cohesion, and there is little doubt that this is of electrical nature. But when the rod is set in motion, the electrical forces inside it must change. For example, each electric charge when put in motion becomes *an electric current*; and the currents will exert magnetic attractions on each other which did not occur in the system at rest. Under the new system of forces the molecules will have to find new positions of equilibrium; they become differently spaced; and it is therefore not surprising that the form of the rod changes. Without going beyond the classical laws of Maxwell we can anticipate theoretically what will be the new equilibrium state of the rod, and it turns out to be contracted to the exact amount required by the Michelson-Morley result.

The contraction of the moving rod ought not to surprise us; it would be much more surprising if the rod were to maintain the same form in spite of the alteration of the electrical forces which determine the spacing of the molecules. But the remarkable thing is that the contraction is only apparent according to the outlook of the solar observer; and we on the earth, who travel with the rod, cannot appreciate it. The fact that the contraction happens to be very small is irrelevant. For convenience suppose that the earth's velocity is 8,000 times faster, so that the contraction amounts to something like a half the original length. We should still fail to notice it in everyday life. Let us say that the direction of the earth's motion is vertically upwards. I turn my arm from horizontal to vertical and it contracts to half its length. No, you cannot convince me I am wrong; I am not afraid of a yard-measure. Bring one and measure my arm; first horizontally, the result is 30 inches; now vertically, the result is 30 – half-inches! Because you must remember that you have turned the scale into the line of the earth's motion so that each inch-division contracts to half an inch. 'But we can see that your arm does not contract. Are we not to trust our eyes?' Certainly *not*, unless you first correct your visual impressions for the contraction of the retina in the vertical direction, and for the effect of our rapid motion on the apparent direction of propagation of the waves of light. You will find, when you calculate these corrections, that they just conceal the contraction. 'But if the contraction takes place, ought one not to feel it happening to the arm?' Not necessarily; I am an observer on the earth, and my feelings like other sense-impressions belong to the geocentric outlook on nature, which Copernicus has persuaded us to abandon.

Take a pair of compasses and twiddle them on a sheet of paper. Is the resulting curve a circle or an ellipse? Copernicus from his standpoint on the sun declares that owing to the FitzGerald contraction the two points drew nearer together when turned in the direction of the earth's orbital motion; hence the curve is flattened into an ellipse. But here I think Ptolemy has a right to be heard; he points out that from the beginning of geometry circles have always been drawn with compasses in this way, and that when the word 'circle' is mentioned every intelligent person understands that this is the curve meant. The same pencil line is in fact a circle in the space of the terrestrial observer and an ellipse in the space of a solar observer. It is at the same time a moving ellipse and a stationary circle. I think that illustrates as well as possible what we mean by **the relativity of space.**

It is sometimes complained that Einstein's conclusion that the frame of space and time is different for observers with different motions tends to make a mystery of a phenomenon which is not after all intrinsically strange. We have seen that it depends on a contraction of moving objects which turns out to be quite in accordance with Maxwell's classical theory. But even if we have succeeded in explaining it to ourselves intelligibly, that does not make the statement any the less true! A new result may often be expressed in various ways; one mode of statement may sound less mysterious; but another mode may show more clearly what will be the consequences in amending and extending our knowledge. It is for the latter reason that we emphasize the relativity of space – that lengths and distances differ according to the observer implied. Distance and duration are the most fundamental terms in physics; velocity, acceleration, force, energy, and so on, all depend on them; and we can scarcely make any statement in physics without direct or indirect reference to them. Surely then we can best indicate the revolutionary consequences of what we have learnt by the statement that distance and duration, and all the physical quantities derived from them, do not as hitherto supposed refer to anything absolute in the external world, but are relative quantities which alter when we pass from one observer to another with different motion. The consequence in physics of the discovery that a yard is not an absolute chunk of space, and that what is a yard for one observer may be eighteen inches for another observer, may be compared with the consequences in economics of the discovery that a pound sterling is not an absolute quantity of wealth, and in certain circumstances may 'really' be seven and six-pence. The theorist may complain that this last statement tends to make a mystery of phenomena of currency which have really an intelligible explanation; but it is a statement which commends itself to the man who has an eye to the practical applications of currency.

Ptolemy on the earth and Copernicus on the sun are both contemplating the same external universe. But their experiences are different, and it is in the process of experiencing events that they become fitted into the frame of space and time – the frame being different according to the local circumstances of the observer who is experiencing them. That, I take it, is Kant's doctrine, 'Space and time are forms of experience.' The frame then is not in the world; it is supplied by the observer and depends on him. And those relations of simplicity, which we seek when we try to obtain a comprehension of how the universe functions, must lie in the events themselves before they have been arbitrarily fitted into the frame. The most we can hope for from any frame is that it will not have distorted the simplicity which was originally present; whilst an ill-chosen frame may play havoc with the natural simplicity of things. We have seen that the simplicity of planetary motions was obscured in Ptolemy's frame, and became apparent in Copernicus's frame. But for ordinary terrestrial phenomena the position is reversed and Ptolemy's frame allows their natural simplicity to become apparent. In Copernicus's frame the most simple phenomena are brought about by highly complicated processes which mutually cancel one another. Ordinary objects contract and expand as they are moved about, and the changes are concealed by an elaborate conspiracy in which all the quantities of nature – electrical, optical, mechanical, gravitational – have joined. In Copernicus's frame we have a great complication of description which has no counterpart in anything occurring in the external world; because the terms of our description refer to the irrelevant process of fitting into the selected frame of space and time. This elaborate Copernican scheme rather reminds one of the schemes of the White Knight –

> But I was thinking of a plan
> To dye one's whiskers green,
> And always use so large a fan
> That they could not be seen.

We do not deny the subtlety and the remarkable efficiency of the plan; but we may be allowed to question whether it is the simplest interpretation of the drab monotony of the face of nature presented to us. The simple fact is that a terrestrial or Ptolemaic frame fits naturally the terrestrial phenomena, and a solar or Copernican frame fits the phenomena of the solar system; but we cannot make one frame serve for both without introducing irrelevant complications.

We go beyond Copernicus nowadays, and are not content with a visit to the sun. Why choose the sun rather than some other star in order to obtain an undistorted view of things ? The astronomer now places himself so as to travel with the centre of gravity of the stellar universe, and is not even then quite satisfied. The physicist dreams of a land of Weissnichtwo, which shall be truly at rest in the ether. We realize the distortion imported into the world of nature by the parochial standpoint from which we observe it, and we try to place ourselves so as to eliminate this distortion – so as to observe

14

that which actually *is*. But it is a vain pursuit. Wherever we pitch our camera, the photograph is necessarily a two-dimensional picture distorted according to the laws of perspective; it is never a true semblance of the building itself.

We must try another plan. I do not think we can ever eliminate altogether the human element in our conception of nature; but we can eliminate a particular human element, namely, this framework of space and time. If our thought must be anthropocentric, it need not be geocentric. Nor are we permanently better off if we merely substitute the space-time frame of some other star or standard of motion. We must leave the frame entirely indeterminate. When we do that, we find that the world common to all observers – in which each observer traces a different space-time frame according to his own outlook – is a world of four dimensions. When we look at any object, say a chair, the impression on our eyes is a two-dimensional picture depending on the position from which we are looking; but we have no difficulty in conceiving of the chair as a solid object, not to be identified with any one of our two-dimensional pictures of it, but giving rise to them all as the position of the observer is varied. We must now realize that this solid chair in three dimensions is itself only an appearance, which changes according to the motion of the observer, and that there is a super-object in four-dimensions, not to be identified with the three-dimensional chair in Ptolemy's scheme, or the same chair in Copernicus's scheme, but giving rise to both these appearances. The synthesis of a three-dimensional chair from a number of flat pictures is easy to us because we are accustomed to assume different positions in rapid succession; indeed our two eyes give us slightly different points of view simultaneously. By sheer necessity our brains have been forced to construct the conception of the solid chair to combine these changing appearances. But we do not vary our motion to any appreciable extent and our brains have not hitherto been called upon to combine the appearances for different motions; thus the effort which we now ask the brain to make is a novel one. That explains why the result seems to transcend our ordinary mode of thought.

The discovery, or one should rather say the rediscovery, of the world of four dimensions is due to Minkowski. Einstein had worked out fully the relations between the frames of space and time for observers with different motions. To the genius of Minkowski we owe the realization that these frames are merely systems of partitions arbitrarily drawn across a four-dimensional world which is common to all observers.

There is a strange delusion that the fourth dimension must be something wholly beyond the conception of the ordinary man, and that only the mathematician can be initiated into its mysteries. It is true that the mathematician has the advantage of understanding the technical machinery for solving the problems which may arise in studying the world of four dimensions; but as regards the conception of the four dimensions of the world his point of view is the same as that of anybody else. Is it supposed that by intense thought he throws himself into some state of trance in which he perceives some hitherto unsuspected direction stretching away at right angles to length, breadth, and thickness? That would not be much use. The world of four dimensions, of which we are now speaking, is perfectly familiar to everybody. It is obvious to every one – even to the mathematician – that the world of solid and permanent *objects* has three dimensions and no more; that objects are arranged in a threefold order, which for any particular individual may be analysed into right-and-left, backwards-and-forwards, up-and-down. But it is no less obvious to every one that the world of *events* is of four dimensions; that events are arranged in a fourfold order, which in the experience of any particular individual will be analysed into right-and-left, backwards- and-forwards, up-and-down, *sooner-and-later*. The subject of our study is external nature, which is a world of events, common to all observers but represented by them differently in their parochial frames of space and time; it is obvious to the most commonplace experience that this absolute world contains a fourfold order.[2]

The news that the events around us form a world of four dimensions is as stale as the news that Queen Anne is dead. The reason why the relativist resurrects this ancient truism is because it is only in this undissected combination of four dimensions that the experiences of all observers meet. In our own experience one dimension is sharply separated from the other three and is distinguished as time; but our experience is solely terrestrial, and if we insist on building the scheme of nature on purely terrestrial experience we are limiting ourselves to the mediaeval geocentric system of the world.

We have been accustomed to regard the enduring world as composed of a continuous succession of instantaneous states, as though the world of events were *stratified*. Each event is supposed to lie in a definite instant or stratum, and the orderly succession of these strata makes up the whole of reality. The instant 'now' represents one such stratum running throughout the universe. Indeed we are accustomed to extend it beyond the universe, and we even use the word 'now' with reference to the existence of those who have passed away from the material world. The investigations of the relativity theory show incontrovertibly that this supposed stratification is an illusion; there is not the slightest evidence for such a view of world-structure. The instantaneous state, which we have hitherto taken to be a nat-

---

[2]The relativity theory does not suggest that there is such a thing in nature as a four-dimensional *space*. The whole object of the recognition of the four-dimensional world is to eliminate the harassing frame of space.

ural stratum in the four-dimensional world of events, is merely an arbitrary partition created by ourselves to correspond with our geo-centric outlook. We can take a differently inclined partition,[3] that is to say, a section which includes on the one side of us events which happened a little while ago and on the other side of us events which have not yet happened; such a farcical combination is in every way equivalent to our so-called instantaneous state, and indeed it *is* an instantaneous state according to the outlook of some non-terrestrial observer with suitably assigned motion.

It is so contrary to our natural prejudices to recognize that the world-wide instant NOW is created by ourselves and has no existence apart from our geocentric outlook, that I will spend a few moments trying to show its artificiality. When I say that I am conscious of an instant NOW, I am only conscious of it in so far as it is HERE – inside me. What then has led me to imagine that there exists a continuation of i.t outside me? It is because I look out on the world and see various events happening 'now', so that I jump to the conclusion that this instant of which I am conscious has to be extended to include them. But that idea is another inheritance from the dark ages, overthrown by Römer in 1675. It is not the events themselves but the sense-impressions to which they give rise which are happening in the instant NOW. So my justification for placing the events outside me in the instants of which I am conscious has entirely disappeared. Unfortunately, however, the crude outlook was not abolished, but patched up; it was found that the immediate difficulties could be met by locating the external events not in the instant of our visual perception of them but in an instant which we had experienced a little time back – allowing, as we say, for the time of propagation of light. Thus our instants were still made to extend through space; but they were carried like partitions among the events by an artificial process of computation, and no longer by immediate intuition. The relativity theory recognizes these *world-wide instants* for what they are – artificial partitions constructed for purposes of calculation. I may add that it in no way tampers with the local *instants* which form the stream of our consciousness;

---

[3]The inclination must not exceed a certain limit. This limiting angle may be regarded as a fundamental constant of the world-structure, and owing to its fundamental character it appears in many kinds of phenomena; for example, it determines the velocity of propagation of light. The instant on the sun which is simultaneous with a given instant on the earth is indeterminate (varying according to the space and time frame employed) but only within a range of 16 minutes. Any event on the sun happening before this 16 minutes is *absolutely* in the past, all observers agreeing on this point; in fact it would be possible for us to have already received a wireless message announcing its occurrence. Events after the 16 minutes are in the *absolute* future. The neutral zone which is (absolutely) neither past nor future becomes proportionately wider as the distance increases; at the nearest fixed star it extends to 8 years, and at the most distant stars yet known it reaches 400,000 years.

it fully recognizes that the chain of events in such a time-succession is a series of an entirely distinctive character from the succession of points along a line in space. Those who suspect that Einstein's theory is playing unjustifiable tricks with time should realize that it leaves entirely untouched that time-succession of which we have intuitive knowledge, and confines itself to overhauling the artificial scheme of time which Römer first introduced into physics.

The study of the four-dimensional world of events gives us a new insight into the processes of nature because it removes the irrelevant stratification in a particular direction – the instantaneous states – which we have so unnecessarily introduced in our customary outlook. When this stratification is ignored we are enabled to see the processes in their simplest aspect, though not, of course, in their most familiar aspect. We must distinguish between simplicity and familiarity; a pig may be most familiar to us in the form of rashers, but the *unstratified* pig is a simpler object of study to the biologist who wishes to understand how the animal functions.

I will conclude this part of the argument with an experimental application which illustrates the power of Einstein's method. Much study has of late been given to electrons moving with very high speeds; for example, the β particles shot off from radioactive substances are negative electrons which sometimes attain speeds of 100,000 miles a second. It is found by experiment that the rapid motion produces an increase of mass of these particles. I want to show that the theory of relativity gives a very simple explanation of just how this increase of mass occurs. But I must first remark that an explanation had been previously given which had generally been accepted as satisfactory. The phenomenon was actually predicted by J. J. Thomson before relativity was thought of; because, assuming that the mass of a β particle is of electrical origin, an application of Maxwell's equations shows that it ought to increase with velocity. But the precise law of increase cannot be predicted on this basis, since various plausible assumptions lead to slightly different results. Moreover, Maxwell's equations are after all only empirical laws, with a mystery of their own; it was a notable advance to connect the change of mass at high speeds with other phenomena whose strangeness has disappeared by long familiarity, but there is still scope for a more far-reaching explanation. Einstein takes us straight to the root of the mystery, and he clears up one point which was misleading, if not actually wrong, in the older explanation. The change of mass does not in any way depend on whether the mass is of electrical origin or not; it arises simply from the fact that mass is a *relative* quantity, depending by its definition on the relative quantities length and time. Let us look at the β particle from its own point of view; it is just an ordinary electron in no way different from any other. 'But it is travelling unusually rapidly?' 'That,' says the electron, 'is a matter of opinion. So far as I am aware I am at rest, if the word "rest" has any meaning. In fact I was just contemplating with amazement *your* extraordinary speed of 100,000 miles a second with which you are shooting past me.' Of course our motion is of no particular concern to the electron, and it will not modify its constitution on our account; so it keeps its mass, radius, electric field, etc., equal to the standard constants applying to electrons in general. These terms are relative, and refer therefore to some particular frame of space and time – clearly the frame appropriate to an electron in self-contemplation, viz. the one with respect to which it is at rest. But this frame is not the usual geocentric frame to which *we* refer quantities such as length, time, and mass; there is a difference of 100,000 miles a second between our station of observation and that of the β particle in self-contemplation. It is a mere matter of geometry to discover what the β particle's lengths and times become when referred to the partitions which we have drawn across the world. But when we calculate the consequential change of mass resulting from the changes of length and time, we find that it should be increased in precisely the proportion indicated by the most refined experiments.

The point is that every electron, at rest or in motion, is a perfectly constant structure; but we distort it by fitting it into the space-time frame appropriate to our own motion with which the electron has no concern. The greater our motion with respect to the electron, the greater will be the distortion. The distortion is not produced by any physical agency at work in the electron; it is a purely subjective distortion depending on our transformation of the reference frame of space and time. This distortion involves a change in our physical description of the electron in terms of mass, shape, size; and in particular the change of mass agrees precisely with that found experimentally.

You see that it is not altogether idle discussing the natural space-time frames for observers moving with huge velocities. We know of no animate observers with these speeds; but we do know of inanimate material objects. Their common resemblance is obscured when we refer them indiscriminately to our irrelevant geocentric frame; we think they have altered their properties, varied in mass, and so on; but the resemblance is restored when we refer each individual to the frame appropriate to it, and so describe them all in comparable terms.

Our measurements of distance in space are found to be subject to certain laws – the laws of geometry. But it has now become impossible to regard the subject of space-geometry as complete in itself. Consider a triangle formed by three points (or events) in the four-dimensional world; if we happen to have drawn our instantaneous strata so that the three points lie in one stratum, then the triangle is a space-triangle and its properties fall within the scope of our classical geometry. But another observer will draw his strata in a different direction, and for him the triangle would be partly in space and partly in time, so that it would not be a fit subject for space-geometry. The subject of geometry is in a desperate condition, because Copernicus and Ptolemy not merely disagree as to the geometry of a configuration; they even disagree as to whether a given configuration is one to which space-geometry is applicable. It is clear that to save it we must extend our geometry so as to include time as well as space. Let me give an illustration of this extension. The terrestrial observer can have a space-triangle (formed by three points or events at the same instant) whose sides he can measure with scales; he can also have a 'time-triangle,' formed by three events on different dates, whose sides he must measure with *clocks*.[4] You all know the law of the space-triangle – that if you measure with a scale from A to B and from B to C the sum of the readings is always greater than the measure from A to C. It is not so well known that there is a precisely analogous law for the time-triangle – that if you measure with a clock from A to B and from B to C the sum of the readings is always *less* than the reading of a clock measuring directly from A to C. In the space-triangle any two sides are together *greater* than the third side; in the time-triangle two sides are together *less* than the third side.[5] Both these laws must be combined in our general geometry of four dimensions, so that it will not be quite so simple a geometry as that to which we are accustomed.[6]

But the point to which I would especially direct attention is this. Evidently the proposition which I have given you about time-triangles cannot be dissociated from the corresponding proposition about space-triangles. When we give up the mediaeval geocentric standpoint, we must recognize that they belong to one geometry, of which our ordinary space-geometry is only a part or projection. But if you examine the proposition about time-triangles, you will see that it is a statement about the behaviour of clocks when they move about, a subject which obviously comes under the heading of mechanics. When we deal with the four-dimensional world we can no longer distinguish between geometry and mechanics. They become the same subject. When we have completely mastered the geometry

---

[4]The three events must not be at the same place since that would give a time-*line* not a triangle. The clock must move so that the two events whose time-distance is to be determined both happen where it is, just as the scale must be directed so that the two points fall on it. You are not allowed to 'bend' the clock, i. e. apply force so as to make it move with other than uniform velocity, any more than you are allowed to bend the scale by applying force.

[5]Of course, it is not true that *any* two sides are less than the third side. A clock, unlike a scale, can only measure in one direction, viz. from past to future, so that the sides $AB + BC$ and AC can be chosen in only one way.

[6]This involves only a comparatively trifling generalization of Euclidean geometry, not to be confused with the 'non-Euclidean' geometry introduced later in the lecture.

20

of the world of events, we shall have inevitably learnt the mechanics of it. That is why Einstein, studying the geometry of the world and discovering that it was strictly non-Euclidean, found that he was at the same time studying the mechanical force of gravitation. And when he had made up his mind which of the possible varieties of non- Euclidean geometry was obeyed, and so settled the laws of the new geometry, the same decision settled the law of gravitation – a law approximating to, but not identical with, the law which Newton had given.

Here a wide vista opens before us. We see that two great divisions of mathematical physics, viz. geometry and mechanics, have met in the four-dimensional world. It is not merely that mechanical problems can be treated by formulae originally belonging to pure geometry; that device has long been in use. Experimental geometry and mechanics actually relate to the same subject-matter; and the young student who discovers experimental laws with ruler and compasses and cardboard figures, and later goes on to pendulums and spring-balances, is developing a single subject which cannot be divided any more than the subject of magnetism can be divided from electricity.

It is through this unification of geometry and mechanics that I should like to approach the problem of gravitation, showing that a field of force is a manifestation of the geometry of space and time. But I fear that that would be too technical; so we will approach it from a different angle.

We have shown that the contemplation of the world from the standpoint of a single observer is liable to distort its simplicity, and we have tried to obtain a juster idea by taking into account and combining other points of view. The more standpoints the better. Let us now consider another point of view, which we have not previously thought about – the point of view of an observer who has tumbled out of an aeroplane and is falling headlong. In many respects his is an ideal situation – temporarily. Unfortunately on *terra firma* we are continually subjected to a very disturbing influence; we undergo a terrific bombardment by the molecules of the ground, which are hammering on the soles of our boots with a total force of some ten stone weight pressing us upwards. Now our bodies are the scientific appliances which we use to make our common observations of the world. I am sure that no physicist would permit any one to enter his laboratory and hammer on his clocks and galvanometers whilst he was observing with them; at any rate he would think it necessary to apply some corrections for the effect of the disturbance. Let us then allow ourselves to fall freely *in vacuo*; then we shall be free from this disturbing bombardment and able to take a much more natural view of what is going on around us.

Whilst falling, we perform the experiment of letting go an apple held in the hand. The apple is now free, but it cannot fall any more than it was falling already; consequently it remains poised in contact with our hand. In our new outlook – in our new frame of space and time – an apple does not drop. There is no mysterious force accelerating it. And remember that this new frame of space and time is the natural frame of a free observer; whereas the old frame, in which the mysterious accelerating force occurred, was the frame of a very much disturbed observer. It is true that when we look down at the earth we see trees and houses rushing up to meet us; but there is no mystery about that. There is an obvious cause for it; plainly they are being propelled upwards from below by that molecular bombardment which I have mentioned. You see that the apple's view of things is simpler than Newton's. Newton had to invent a mysterious force dragging the apple down; the apple observes only a familiar physical agency propelling Newton up.

It is not my purpose to emphasize unduly the superiority of the apple's view over Newton's, but rather to regard both on an equal footing. I have perhaps been a little unfair to Newton. His position on the surface of the earth was unfortunate, but he would have been perfectly content to be at the centre of the earth, where he could have remained without support, i. e. without disturbance by molecular bombardment. From there he would still have observed the well-known acceleration of the apple; and the apple would have observed a corresponding acceleration of Newton without any molecular bombardment causing it. From either point of view there is a mysterious agent at work. How shall we picture to ourselves this agent? Shall we picture it as a force – a tug of some kind? But if so, to which of them is the tug applied? If we take the standpoint of Newton the tug is applied to the apple, if the standpoint of the apple the tug is applied to Newton; so that in our synthesis of all standpoints we cannot decide which is being tugged, and the picture of gravitation as a tugging agent becomes impossible. Einstein replaces it by a different picture, which we shall perhaps better understand if we compare it with a very similar revolution of scientific thought which occurred long ago.

The ancients believed that the earth was flat. The small portion of its surface with which they were chiefly concerned could be represented without serious distortion on a flat map. As more distant countries were added, it would be natural to think that they also could be included in the flat map. You have all seen such maps of the world, e. g. Mercator's projection, and you will remember how Greenland appears enormously exaggerated in size. Now those who adhered to the flat-earth theory must hold that the flat map gives the *true* size of Greenland. How then would they explain that travellers in that country reported that the distances were much shorter? They would, I suppose, invent a theory that a demon resided in that country who helped travellers on their way, making the journeys appear much shorter than they 'really' were. No doubt the scientists would preserve their self-respect by using some Graeco-Latin polysyllable instead of the word 'demon', but that must not disguise from us the fact that they were appealing to a *deus ex machina*. The name demon is rather suitable, however, because he has the impish characteristic that we cannot pin him down to any particular locality. We might equally well start our flat map with its centre in Greenland; then it would be found that journeys there were quite normal, and that the activities of the demon were disturbing travellers in Europe. We now recognize that the true explanation is that the earth's surface is curved; and the demoniacal complications appeared because we were forcing the earth's surface into an inappropriate flat frame which distorts the simplicity of things.

What has happened in the case of the earth has happened also in the case of the world, and a similar revolution of thought is needed. An observer, say at the centre of the earth, finds that there is a frame of space and time – a flat or Euclidean frame – in which he can locate things happening in his neighbourhood without distorting their natural simplicity. There is no gravitation, no tendency of bodies to fall, so long as the observer confines his observations to his immediate neighbourhood. He extends this frame of space and time to greater distances, and ultimately to the earth's surface where he encounters the phenomenon of falling apples. This new phenomenon must be accounted for, so he invents a *deus ex machina* which he calls gravitation to whose activities the disturbance is attributed. But we have seen that we may just as well start with the falling apple. It has a flat frame of space and time into which phenomena in its neighbourhood fit without distortion; and from its point of view bodies near it do not undergo any acceleration. But when it extends this frame farther afield, the simplicity is lost; and it too has to postulate the demon force of gravitation existing in distant parts, and for example causing undisturbed objects at the centre of the earth to fall towards it. As we change from one observer to another – from one flat space-time frame to another – so we have to change the region of activity of this demon. Is not the solution now apparent? The demon is simply the complication which arises when we force the world into a flat Euclidean space-time frame into which it does not fit without distortion. It does not fit the frame, because *it is not a Euclidean or flat world.* Admit a curvature of the world and the mysterious disturbance disappears. Einstein has exorcized the demon.

Einstein, recognizing that in the phenomena of gravitation he was not dealing with a 'tug' but with a curvature of the world, had to reconsider the law of gravitation. He could not make any possible law of curvature correspond exactly with the previously assumed law of tugging. Thus he was led to propound a new law of gravitation – a law which in most practical cases differs very little from that of Newton, although it has an essentially different foundation. I need not here dwell on the very remarkable way in which Einstein's emendation of the law of gravitation has been confirmed both by the anomalous secular change in the orbit of the planet Mercury, and by the observed displacement of the stars near the sun during the total eclipse of 1919. I might, however, remind you that in the latter observation the point at issue between Newton's and Einstein's theory was not the *existence* of a deflexion of light-rays passing near the sun but the amount of the deflexion, Einstein predicting twice the deflexion possible on the Newtonian theory. The larger deflexion was quantitatively confirmed by the eclipse observations. Einstein's main achievement is a new law, not a new explanation, of gravitation. He attributes the gravitation of massive bodies to a curvature of the world in the region surrounding them and so throws a flood of light on the whole problem; but he is not primarily concerned to explain how material bodies produce (or are associated with) this curvature of the world around them, nor how this curvature is made subject to a law. Although it would be an entire misunderstanding of Einstein's attitude in propounding the general relativity theory to regard it as a search for an explanation of gravitation, nevertheless I think that the further following up of his ideas has led to a genuine explanation as complete as could be desired. But I am not going to give you the explanation in this lecture; sometimes an explanation requires a great deal of explaining.[7]

---

[7]The following brief outline will give a hint of the nature of the explanation. Einstein's law of gravitation is usually expressed as a set of ten very lengthy differential equations; these equations are exactly equivalent to the geometrical statement that 'the radius of spherical curvature of any 3-dimensional section of the 4-dimensional world is a universal constant length, the same for all points of the world and for all directions of the section'. The law therefore implies that the world has a certain type of homogeneity and isotropy (not, however, the *complete* homogeneity and isotropy of a sphere). To explain the law of gravitation and the phenomena governed by it, we have to explain how this isotropy and homogeneity is secured. Our explanation is that the homogeneity and isotropy is not initially in the external world, but *in the measurements which we make of it*. It is introduced in all our operations of measurement, because the appliances which we use for measurement are themselves part of the world. In the earlier part of this lecture we saw that the contraction of the arm turned from horizontal to vertical is not detected by measurements made with a yard-measure which shares the contraction; in the same way any anisotropy of the world does not appear in measurements of it by appliances which, being part of the world, share the same anisotropy. The law of gravitation therefore arises from the fact that a certain type of non-homogeneity and non-isotropy of the world cannot come into observational experience, because it is necessarily eliminated in all observations and measurements made with material appliances. The orderly phenomena of gravitation are due to the *absence* of certain conceivable effects. We have been trying to find a key to the mystery; but the secret of the lock lies not in the key but in the wards.

I think that we can without mathematics form a general idea of why Einstein found it necessary to amend Newton's law of gravitation. Let us return to the illustration of the pig, and imagine that we wish to discover the law governing the distribution of fat and lean in the animal. From the breakfast-table standpoint a plausible type of law would be that half of each rasher is fat and the other half lean; and if this turned out to be confirmed very approximately by observation we might well imagine that we had discovered the exact law of porcine structure. But the case is altered if we give up the breakfast-table standpoint and contemplate the animal in a more general way, remembering that he has not been designed with any particular reference to the series of rashers into which our grocer has chosen to slice him. We must now look for a different type of law altogether. Two possibilities may arise. We may find that our proposed law, although expressed in breakfast-table parlance, is nevertheless equivalent to a possible biological law; it may be immediately capable of translation into a more general statement which makes no reference to a particular stratification. But on the other hand, it may happen that the suggested law cannot be freed from this reference to a particular system of slicing. In that case we can only regard it as approximate, perhaps holding fairly well for the slices of which we have most experience but becoming less and less accurate in the more tortuous parts of the animal. Both these cases are illustrated in Einstein's modifications of classical theory. Newton's law of gravitation explicitly refers to a space-time frame and therefore to a world stratified into instantaneous states. It proves to be impossible to free it from this reference to a particular stratification without modifying it. In fact if the crucial astronomical observations had shown that Newton's law and not Einstein's was the exact law of gravitation, this would have been evidence of a real stratification of the structure of the world – a stratification revealed by no other phenomena. Einstein's law is the simpler law because it is consistent with what we now know of the general plan of world-structure; Newton's law could only be made possible by introducing a novel and specialized feature – a stratified arrangement of structure – which is not revealed in any other phenomena.

Maxwell's laws of electromagnetism afford an example of the other type. These, it is true, are stated as relating to the particular slices of the world of events, which are served up to us like rashers instant by instant. But they can be restated, without alteration of effect, in a form making no reference to slices. This is a very remarkable property of Maxwell's equations which was quite unknown at the time they were first put forward. It was brought to light much later by the researches of Larmor and Lorentz. In consequence of this Einstein is able to take over the whole classical theory of electromagnetism unaltered; he restates it so as to show how it applies generally and is not bound up with the purely terrestrial point of view, but he does not amend the laws. He metes out different treatment to the gravitational laws and electromagnetic laws, because he finds the latter already adapted to his scheme.

If I have succeeded in my object, you will have realized that the present revolution of scientific thought follows in natural sequence on the great revolutions at earlier epochs in the history of science. Einstein's special theory of relativity, which explains the indeterminateness of the frame of space and time, crowns the work of Copernicus who first led us to give up our insistence on a geocentric outlook on nature; Einstein's general theory of relativity, which reveals the curvature or non-Euclidean geometry of space and time, carries forward the rudimentary thought of those earlier astronomers who first contemplated the possibility that their existence lay on something which was not flat. These earlier revolutions are still a source of perplexity in childhood, which we soon outgrow; and a time will come when Einstein's amazing revelations have likewise sunk into the commonplaces of educated thought.

To free our thought from the fetters of space and time is an aspiration of the poet and the mystic, viewed somewhat coldly by the scientist who has too good reason to fear the confusion of loose ideas likely to ensue. If others have had a suspicion of the end to be desired, it has been left to Einstein to show the way to rid ourselves of these 'terrestrial adhesions to thought'. And in removing our fetters he leaves us, not (as might have been feared) vague generalities for the ecstatic contemplation of the mystic, but a precise scheme of world-structure to engage the mathematical physicist.

THE RELATIVITY OF TIME

A. S. Eddington, The Relativity of Time, *Nature* **106**, 802-804 (17 February 1921)

The philosopher discusses the significance of time; the astronomer measures time. The astronomer goes confidently about his business and does not think of asking the philosopher what exactly is this thing he is supposed to be measuring; nor does the philosopher always stop to consider whether time in his speculations is identical with the time which the world humbly accepts from the astronomer. In these circumstances it is not surprising that some confusion should have arisen.

In many globular clusters there are stars which oscillate in intrinsic brightness; let us select two such stars from different clusters and invite all the astronomers in the universe to measure the true interval of time between the moments of maximum light of the two stars. They must, of course, make whatever measurements and calculations they consider necessary to allow for the finite velocity of light. It may easily happen that the astronomers on Arcturus report that the two maxima were simultaneous; whereas those on the earth report an interval of *ten years* between the same two maxima. There is here no question of observational error; the recognised terrestrial method necessarily gives a discordant result when on Arcturus, owing to its different motion.

Our first impulse is to blame the astronomers. Evidently they are not giving us the true time-interval; and now that they are informed of the discordance they ought to give up their out-of-date procedure. But the astronomers reply: "Tell us, then, how we ought to find this 'true time'. By what characteristics are we to recognize it?" No answer has been given. Michelson and others sought in vain for an answer; for if our velocity through the aether could be defined, it would single out one universal system of time-measurement which might reasonably (if somewhat arbitrary) be called true. Meanwhile the phrase *true time* is a "meaningless noise." It is idle to contest with those who hold that the thing exists and ought to be regarded. "Who would give a bird the lie, though he cry 'Cuckoo' never so?"

The direction of Northampton measured by astronomers at Cambridge is due west; measured by astronomers at Greenwich it is north-west. It is no use to tell them that they must adopt a different plan, and find a "true direction" of Northampton which does not show these discordances. They reply: "We are perfectly aware that there must be discordances, as you call them; but that is in the nature of a relative property like direction; as for this true direction which shall be the same from all stations, we have no idea what you are talking about."

The time determined by astronomers and in general use is thus a fictitious time, or, in the usual phrase, it is *relative* to terrestrial observers. Similarly it has been found that extension in space is also relative. When the Copernican theory led to the abandonment of the geocentric view of the universe, the revolution did not go far enough; it was thought that we could pass to the heliocentric outlook by merely allowing for what in pure geometry would be called a change of origin. Actually a more profound transformation is necessary. For example, the Michelson-Morley experiment is a terrestrial experiment, but its theory is treated from a heliocentric point of view; that is to say, account is taken of the varying orbital motion of the earth; it finishes a proof of the famous FitzGerald contraction, and much ingenuity has been spent on an electrical explanation of this curious property of matter. Einstein's theory waves this aside with the remark: "Of course, your results appear strange when you describe the apparatus in terms of a space and time which do not belong to it. Your electromagnetic discussion is no doubt valid, but it is leading you away from the root of the matter; the immediate explanation lies in the difference between the heliocentric and geocentric space and time systems."

It was shown by Minkowski that all these fictitious spaces and times can be united in a single continuum of four dimensions. The question is often raised whether this four-dimensional space-time is real, or merely a mathematical construction; perhaps it is sufficient to reply that it can at any rate not be less real than the fictitious space and time which it supplants. Terrestrial observers divide the four-dimensional world into a series of sections or thin sheets (representing space) piled in an order which signifies time; in other words, the enduring universe is analyzed into a succession of instantaneous states. But this division is purely geometrical. The physical structure of the enduring world is not laminated in this way; and there is nothing to prevent another observer drawing his geometrical sections in a different direction. In fact, he will do so if his motion differs from ours. Now it may seem that we have been paying too much deference to the astronomers: "After all, they did not discover time. Time is something of which we are immediately conscious." I venture to differ and to suggest that (subject to certain reservations) time as now understood *was* discovered by an astronomer – Römer. By our sense of vision it appears to us that we are present at events far distant from us, so that they seem to occur in instants of which we are immediately conscious. Römer's discovery of the finite velocity of light has forced us to abandon that view; we still like to think of *world-wide* instants, but the location of distant events among them is a matter of hypothetical calculations, not of perception. Since Römer, time has become a mathematical construction devised to give the least disturbance to the old illusion that the instants in our consciousness are world-time.

Without using any external sense, we are conscious of the flight of time. This, however, is not a succession of world-wide states, but a succession of events at one place – not a pile of sheets, but a chain of points. Common-sense demands that this time-succession should be essentially different from the space-succession of points along a line. The preservation of a fundamental distinction between timelike succession and spacelike succession is essential in any acceptable theory. Thus in the four-dimensional world we recognize that there are two types of ordered succession of events which have no common measure; type A is like the succession of instants in our minds, and type B is the relation of order along a line in space. Proceeding from the instant "here-now", I can divide the regions of the world into two zones, according as they are reached by a succession of type A (my absolute past and future), or of type B (my absolute "elsewhere"). This scheme of structure is very different from the supposed laminated structure of the older view. Since we believe that this distinction of types A and B corresponds to something in the actual structure of the world, it is likely to determine the various natural phenomena that are observed. Thus it determines the propagation of light, since it is found that the line of a light-pulse is always on the boundary between the two zones above-mentioned. More important still, a particle of matter is a structure which can occupy a chain of points only of type A. Since we are limited by our material bodies, it must be this type of succession which we immediately experience; we are aware of the existence of the other type only by deduction from the indications of our external senses.

Objection is sometimes raised to the extravagantly important part taken by light-signals and light-propagation in Einstein's discussion of space and time. But Einstein did not invent a space and time depending on light-signals; he pointed out that the space and time already in general use depended on light-signals and equivalent processes, and proceeded to show the consequences of this. Turning from fictitious space and time to the absolute four-dimensional world, we still find the velocity of light playing a very prominent part. It is scarcely necessary to offer any excuse for this. Whether the substratum of phenomena is called *aether* or *world* or *space-time*, one requirement of its structure is that it should propagate light with this velocity.

The resolution of the four-dimensional continuum into a succession of instantaneous spaces is not dictated by anything in the structure of the continuum. Nevertheless, it is convenient, and corresponds approximately to our practical out-look on the world; and it is rarely necessary to go back to the undivided world. We have to go back to the undivided world when a comparison is made between the phenomena experienced by observers with different motions, who make the resolution in different directions. Moreover, a world-wide resolution into a space and time with the familiar properties is possible only when the continuum satisfies certain conditions. Are these conditions rigorously satisfied? They are not; that is Einstein's second great discovery. It is no more possible to divide the universe in this way than to divide the whole sky into squares. We have tried to make the division, and it has failed; and to cover up the consequences of the failure we have introduced an almost supernatural agency – gravitation. When we cease to strive after this impossibility - a mode of division which there was never any adequate reason for believing to be possible – gravitation as a separate agency becomes unnecessary. Our concern here is with the bearing of this result on time. The relative time for an observer is a construction extended by astronomers throughout the universe according to mathematical rules; but these rules break down in a region disturbed by the proximity of heavy matter, and cannot be fulfilled accurately. We can preserve our time-partitions only by making up fresh rules as we require them. The local time for a particular observer is always definite, and is the physical representation of the flight of instants of which he is immediately aware; the extended mesh-work of co-ordinates radiating from this is drawn so as to conform roughly to certain rules – so as not to violate too grossly certain requirements which the untutored mind thought necessary at one time. Subject to this, time is merely one of four co-ordinates, and its exact definition is arbitrary.

To sum up, world-wide time is a mathematical system of location of events according to rules which on examination can only be regarded as arbitrary; it has not any structural – and still less any metaphysical – significance. Local time, which for animate beings corresponds to the immediate time-sense, is a type of linear succession of events distinct from a pure spacelike succession; and this distinction is fully recognized in the relativity theory of the world.

THE MEANING OF MATTER AND THE LAWS OF NATURE ACCORDING TO THE THEORY OF RELATIVITY

A. S. Eddington, The Meaning of Matter and the Laws of Nature according to the Theory of Relativity, *Mind* **29** (114) 145 – 158 (1920)

The theory of relativity has introduced into physics new conceptions of time and space, which have aroused widespread interest. Less attention has been paid to the position of matter in the new theory; but a natural interpretation suggests a view of the nature of matter, which is in some respects novel and is more precise than the theories hitherto current. It is perhaps a commonplace that, whatever may be the true nature of matter, it is the *mind* which from the crude substratum constructs the familiar picture of a substantial world around us. On the present theory we seem able to discern something of the motives of the mind in selecting and endowing with substantiality one particular quality of the external world, and to see that practically no other choice was possible for a rational mind. It will appear in the discussion that many of the best-known laws of physics are not inherent in the external world, but were automatically imposed by mind when it made the selection.

Probably the views here reached accord in a general way with some recognised philosophical theory; but it will be of interest to show how they are approached from the physical side. I must crave indulgence for the very imperfect expression of my ideas, being on the one hand debarred from using the conventions and terminology of mathematics, and on the other hand insufficiently expert to use the technical terms of philosophy.

It is convenient first to make some remarks on the general nature of physical theories. We believe that the ordinary objects of experience are very complex; in order to understand their mutual relations and to "explain" the phenomena, they must be resolved into simpler elements. Whilst it is a reasonable procedure to explain the complex in terms of the simple, this necessarily involves the paradox of explaining the familiar in terms of the unfamiliar. Thus the ultimate concepts of physics are of a nature which must be left undefined; we may describe how they behave, but we cannot state what they *are* in any terms with which the mind is acquainted. The entities which appear in physical theories fall into three categories. We take for illustration the electromagnetic theory of light. There is first the aether. The word brings before the mind the idea of a limitless ocean pervading space; but during the last century, all the properties which would make the aether akin to any known fluid have had to be abandoned one by one. At the present time it would seem that the only property it possesses in common with a material ocean is that of being three-dimensional and even this is now challenged by the relativity theory. To describe the nature (as distinct from the *properties*) of such a medium in terms familiar to the mind is impossible. Further, the aether is not in itself a subject for physical measurement. Secondly, there are quantities like electric and magnetic force; their nature is undefinable but their intensity can be measured by practical experiment. It is fundamental in the theory of relativity that anything measurable must necessarily be of the nature of a *relation* between two or more constituents of the external world; accordingly, we call objects of this second class relations. Thirdly, we have light, an object of experience; it is something common to our mental picture of the universe and to the analytical world of physics. The three classes are accordingly: (1) elementary analytical concepts, undefinable and unmeasurable; (2) relations, undefinable but measurable; (3) objects of experience, which are definable.

There is no particular awkwardness in developing a mathematical theory in which the elementary constituents are undefined. But it is desirable that at some stage in the discussion we should get to know what it is we are talking about; and this is achieved when we can identify one of the complex combinations of our undefinables with some object of experience recognised by the mind. Strange as it may seem, it is quite easy to overlook this necessity.

An objection may be raised here. Do not the things which can be measured – time, mass, electric force, etc. – come within experience? And may we not be satisfied when we reach the stage of dealing with things which can be measured? The physicist is satisfied, and rightly so; but then he is not usually occupied with evolving a complete scheme of things. Now all measures are made with the help of undoubted objects of experience – clocks, scales, galvanometers, etc. – and if we are to make a complete theory, to understand how the galvanometer measures an electric current, we must first learn what a galvanometer is in terms of electric currents and the other simpler concepts of the theory. In other words the theory must be developed until it reaches some combination which can be identified as a galvanometer.

There are, in fact, a number of possible sites for a bridge between the analytical theory and the phenomena of perception. As has been said, the physicist commonly makes the connexion through things that are measured experimentally. Another alternative is to carry on the analytical development of the external world to the point at which it meets mind in the nerve-centres of the brain. In this paper I have taken the middle course of making the connexion through the everyday world which we see and feel around us. I regard the objects of this world as immediately recognisable to the mind – they are our definables – so that it is here that the bridge is most naturally made. We can to a certain extent *think forward* to electric currents, or *think backward* to mental processes; but it is more in accordance with the mathematical ideal to cross the bridge at this point, and carry on any further investigations in the analytical world rather than in the perceptual world.

In the relativity theory of nature the elementary analytical concept is the "point-event." In ordinary language a point-event is an instant of time at a particular point in space; but this is only one aspect of the point-event, and must not be taken as a definition, because the space and time of experience are derived concepts of considerable complexity. From what has already been said, it will be understood that the point-event is necessarily undefinable and its nature is outside the range of human understanding. The aggregate of all the point-events is called the "World";[1] and we postulate that this aggregate is four-dimensional. Pure mathematicians have, I believe, evolved a logical definition of the property implied by the term four-dimensional without appealing to intuitive notions of space and time; and it results that a particular point-event can be specified by the values of four variables or co-ordinates, which in practice are usually taken as three co-ordinates of space and one of time. Between any two neighbouring point-events there is a certain relation known as the "interval" between them. The relation is a quantitative one and can be assigned a numerical value. The term "interval" must not be taken as any guide to the real nature of the relation, which is beyond our power to conceive. The name refers not to its nature but to certain of its properties (ascertained later), which are those of a geometrical interval in a very extended mathematical sense – extended, because, for example, when the interval vanishes the two point-events are not necessarily identical.[2] The interval is not quite so transcendental as the point-event, because we are able to measure an interval practically with scales and clocks; but this is an anticipation of results which are only reached at a much later stage. Accordingly at present we are still pursuing a purely analyti-

---

[1]The capital letter will show when the word is used in this technical sense.

[2]Point-events may be compared to straight lines in three-dimensional space, and the interval to the shortest distance between them. When the shortest distance vanishes the two lines intersect but are not necessarily coincident.

cal development which has not as yet been connected with anything in nature which can be perceived or measured.

What we have here called the World might perhaps have been legitimately called the aether; at least it is the universal substratum of things which the relativity theory has given us in place of the aether. But the aether in physical theories has been gradually changing its character as science has developed, and perhaps this latest change from a three-dimensional to a four-dimensional aggregate is sufficiently fundamental to justify a new name.

Consider a small portion of the World. It consists of a large (possibly infinite) number of point-events, between every two of which an interval exists. If we are given the intervals between a point-event A and a sufficient number of other point-events, and also between B and the same point-events, can we calculate what will be the interval between A and B? In ordinary geometry there are rules for doing this; but in the present case, knowing nothing of the nature of the relation signified by the word interval, clearly we cannot predict any law *a priori*. There may be in any small region some law for calculating the interval AB, which need not be the same in all parts of the World. Whether this is so or not, and even if the individual intervals are entirely arbitrary and discontinuous, we may take the rule which best represents the average for the region; and the coarse measures of physics appreciate only the average. This rule, or average rule, of connexion of intervals expresses a quality of the World at the region considered, and may reasonably vary from region to region. One part of the World differs from another part – an intrinsic absolute difference – and this on our theory is the starting point for the infinite variety of nature.

An example may help to make this clear. I deliberately choose a non-geometrical example, because we must try to get rid of the obsession that the interval-relation is something geometrical. Compare the point-events to persons, and the intervals to the degree of acquaintance between them. Given the degree of acquaintance between A and C and between B and C, there is no rule for determining the degree of acquaintance between A and B. But a statistician might determine in any community the average rule, or "correlation," between the mutual acquaintance of two individuals, and their acquaintance with a third individual; if A and B know C, it increases the probability of their knowing one another. The correlation may be higher in some communities than in others, and so measure intrinsic differences between communities.

The mathematician measures this quality of the World by a set of coefficients, denoted individually by g_{11}, g_{12}, etc., up to g_{44}, and collectively by $g_{\mu\nu}$. But $g_{\mu\nu}$, besides containing the measure of this absolute quality, contains something else – physical time and space, which we now believe are not intrinsic qualities of the world. Probably the philosopher and the physicist attach somewhat different meanings to time and space; to the former it is the seat of events, to the latter it is in addition the seat of measurement. Philosophical space-time has been implicitly introduced in postulating the World to be four-dimensional; but it is a long step from this to the partitioned space and time of the physicist which serves as a reticule for his measurements. In order then to give definite values to $g_{\mu\nu}$, we have first to choose a system of co-ordinates, i.e., to define a particular way of partitioning space and time; and at the present stage we are not in a position to do this. The way out of the dilemma is to continue the analysis, leaving the space and time undetermined, but making sure that our results will apply whatever system of measuring space and time we ultimately decide to adopt. Fortunately a remarkable calculus has been invented by pure mathematicians for an entirely different purpose, which enables us to pursue the analytical development leaving the co-ordinates entirely undefined.

By considering the variation of $g_{\mu\nu}$ from point to point its gradient and the gradient of the gradient, other more complex characters of a region are obtained. But these involve the undetermined space and time, and our object is rather to refine out from space and time those things which are the intrinsic qualities of World. By an exceedingly complicated combination of these operations, we arrive at a set of quantities called $G_{\mu\nu}$, which serve our purpose.[3] It must be remarked that a complicated mathematical formula may express a comparatively simple idea; for example, the formula for the curvature of a surface is by no means simple, yet everyone can form an idea of the property which it expresses. It is true that the physical conception measured by $G_{\mu\nu}$ is scarcely intelligible to us, but a being capable of conceiving five dimensions would grasp it more easily.

The quantity $G_{\mu\nu}$ plays a fundamental part in Einstein's generalised relativity theory, which asserts as a law of nature that in empty space

$$G_{\mu\nu} - \frac{1}{2}g_{\mu\nu}\, G = 0.^4$$

This is in fact the new law of gravitation, which in all ordinary cases agrees approximately with the Newtonian law of the inverse square, but in addition accounts for the celebrated astronomical discordance

---

[3]Things like $g_{\mu\nu}$ and $G_{\mu\nu}$ (called tensors) occupy a position intermediate between intrinsic qualities of the World, and qualities which involve space and time haphazardly. The *vanishing* of a tensor does actually denote an intrinsic condition quite independent of time and space, and the equality of two tensors in the same region is also an absolute relation. It is for this reason that $G_{\mu\nu}$ (the simplest tensor after $g_{\mu\nu}$) attracts our attention.

[4]G is an abbreviation for a complicated combination of $g_{\mu\nu}$ and $G_{\mu\nu}$. The whole of the left side is a tensor, and therefore, although it does not measure an intrinsic quality of the World, its *vanishing* (expressed by the equation) denotes an intrinsic condition. (See previous footnote.)

of the motion of the perihelion of Mercury. Unlike the Newtonian law, however, it does not presuppose any particular mode of measuring space and time, and it is for that reason especially that it commends itself to those who have a bias in favour of the relativity theory. It expresses a relation between the intrinsic properties of adjacent portions of the World, and not (like the Newtonian law) a relation between these properties and some extraneous space and time.

When matter is present the law is modified by the addition of a term $T_{\mu\nu}$ which is compounded from the density, momentum, stress, and energy of the matter present. The new term is a tensor, and accordingly the equation is still independent of space and time. The equation now reads

$$G_{\mu\nu} - \frac{1}{2}g_{\mu\nu}\,G = -8\,\pi\,T_{\mu\nu}.$$

I suppose that the usual view of these equations is that the first of them expresses some law inherent in the continuum – that the point-events are forced by some natural necessity to arrange themselves so that their relations accord with this law. And when matter intrudes, it disturbs the linkages and causes a rearrangement to the extent indicated by the second equation.

But I think there is something incongruous in introducing an object of experience (matter) as a foreign body disturbing the domestic arrangements of the analytical concepts from which we have been building a theory of nature. It leads to a kind of dualism. What should we think of a chemical theory which, instead of analysing matter into atoms, postulated the existence of non-material atoms in addition to continuous matter and then proceeded to discover laws of nature connecting the behaviour of matter with that of the non-material atoms? There is a redundancy, and whenever we have an unnecessary multiplication of entities we are liable to find spurious laws of nature which are in reality only identifications. The result that the velocity of light is the same as that of electric waves does not determine any law of the aether, but merely the identification of light with electric waves.

We prefer therefore to take another view of the equations of Einstein. The vanishing of the left-hand side in any region denotes a definite and absolute condition of the World in that region; and, if Einstein's theory is true, that condition is common to all parts of the world which are empty of matter. Up to the present we have had no indication of what impression, if any, that condition of the World would make on our senses. *I suggest that it gives us the perception of emptiness.* The left-side of the equation is composed solely of analytical quantities which have not been defined; at some time or other, and preferably at the earliest possible stage in our synthesis, we have to identify the symbols of theory with things familiar to experience – in short, to learn what we are talking about. This is our opportunity. Mind surveying the external world passes over unnoticed many of the differences of quality which from the mathematical standpoint are most elementary; it has developed no faculty for perceiving the quality measured by $g_{\mu\nu}$; but we have now arrived in our discussion at a quality which mind takes cognisance of and recognises under the name of "emptiness." Einstein's law of gravitation is not a law of nature but a definition – the definition of a vacuum.

Similarly when $G_{\mu\nu} - \frac{1}{2}g_{\mu\nu}\,G$ does not vanish, the corresponding property of the world is perceived by us as a distribution of matter. Our second equation teaches us what density and state of motion of matter is the perceptual equivalent of any particular value of this world-property. This again is not a law inherent in the external world, but merely describes how the hitherto undefinable quality measured by the left-hand side of the equation is appreciated by the human mind. Matter does not cause an unevenness in the gravitational field; the unevenness of the field *is* matter.

It may be worth while to turn aside for a moment to point out why the meaning of these equations has been obscured in the usual presentation of the relativity theory. The general course is to start with the "interval" as something immediately measurable with scales and clocks; accordingly $G_{\mu\nu}$ is measurable practically, and the equations are of the type normally encountered in physics in which all the quantities involved are measurable. But in a strict analytical development the introduction of scales and clocks before the introduction of matter is – to say the least of it – an inconvenient proceeding. Thus in our development $G_{\mu\nu}$ is not merely of unknown nature but unmeasurable. The equations therefore connect the familiar and measurable quantities on the right with the hitherto unfamiliar and unmeasurable quantities on the left, and have no value except as definitions.

Our contention, that the introduction of matter as a foreign entity in the gravitational field is superfluous, is so fundamental in what follows that at the risk of repetition we must endeavour to make plain the position taken up. How any physical phenomenon can produce a sensation in the mind must be a great mystery; and it would be difficult to say that any theory of the nature of matter makes our perception of it less or more easily understood. But those who are accustomed to regard the $g_{\mu\nu}$ as coefficients defining the geometry of space may well deem it altogether too fantastic that any combination of these quantities could create a sensation in the mind. But we have seen that the $g_{\mu\nu}$ are undefinables, and so we may attribute to them whatever nature we may conceive as best fitted to affect the mind; their geometrical interpretation is incidental, and is due to the fact that natural geometry depends on observations of the behaviour of matter and therefore ultimately on the behaviour of the $g_{\mu\nu}$. Granting then that a brain constituted of $G_{\mu\nu} - \frac{1}{2}g_{\mu\nu}\,G$ is at least as capable of being the seat of sensation as any other conceivable structure, there is no occasion to introduce any other kind of substance. We do not suppose that a ray of light is a rod which causes the electro-magnetic force to oscillate along its path; the electro-magnetic oscillations constitute the ray of light. We do not suppose that heat is a fluid which causes violent motions of the molecules of a body; the motions constitute heat. So too, we need not suppose that matter is a substance which causes irregularities in the gravitational field; the irregularities are matter. We shall show presently that matter thus defined satisfies the well-known laws of mechanics.

According to this view matter can scarcely be said to exist apart from mind. Matter is but one of a thousand relations between the constituents of the World, and it will be our task to show why one particular relation has a special value for the mind. It need not surprise us that mind appreciates a particular relation rather than the external entities themselves; it is but an instance of the peculiarity that mind sees not the *paint* but the *picture*.

We have thus arrived at a definition of matter in terms of the analytical concepts and their relations. And it must be remarked that matter and the motion of matter have been defined separately. When we have fixed on any arbitrary way of measuring space and time, the different components of the tensor $T_{\mu\nu}$, give separately the density, momentum, and other combinations of the mass and velocity of matter. In practice we detect the motion of a body by noticing that the body has disappeared from one point of space and an apparently identical body has appeared at a neighbouring point. But as here brought in motion has nothing to do with this property. The analytical introduction of motion is rather curious. It is the ratio of two of the components of the World-property $G_{\mu\nu} - \frac{1}{2}g_{\mu\nu}\,G$. We have thus a definition of motion which does not involve the elusive idea of permanent identity of particular particles of matter; nor does it involve the definition of a particular way of measuring space and time, but rather we are able to proceed from it to introduce the partitioned space and time of physics.

Now the expression $G_{\mu\nu} - \frac{1}{2}g_{\mu\nu}\,G$ has a remarkable property known as the property of *conservation*. This property is simply a mathematical identity due to the way in which the expression has been built up from the simpler elements $g_{\mu\nu}$. It results from this property that, *provided we measure space and time in one of a certain limited number of ways*, matter will be permanent; for every particle which disappears at any point of space a corresponding mass will appear at a neighbouring point (conservation of mass). Further, the velocity of matter as introduced in the previous paragraph will agree with the velocity measured in the ordinary way; and this provides the basis of practical methods of defining the space and time here required. Finally, momentum and energy will obey the law of conservation.

These extensive results are in no sense laws of nature; they must hold in any imaginary world just as they do in the actual world. Or if $g_{\mu\nu}$ referred to relations in a human community instead of to intervals of point-events, the same laws must still hold. To predict these laws we need to know nothing about the properties of the constituents of the external world; all that we need to know is, under what names will mind recognise the things which obey the laws?

For some unknown reason the mind appears to have a predilection for living in a more or less permanent universe. The idea of reality is at least closely associated with the idea of permanence. And so the mind has picked out from the external World a universe built from permanent elements (matter), and it is pleased to regard this as the real world. This, we have seen, involves a specialised way of measuring space and time; and so compelling is the desire for permanence that we have adopted this special space and time instinctively and find it hard to realise it is not the only one. I think this is the origin of the singling out of our familiar space and time from the many possible ways of resolving a four-dimensional continuum.

May we not go further? Why is it that of all the properties distinguishing different parts of the World, only one, and that a rather complex one, is perceived by us as substantial? Imagine an embryo mind surveying the external World without form and void – void because as yet mind has not made the final decision "Let *this* be matter." It is at the parting of the ways, uncertain with what feature of this cosmos to develop faculties of recognition. But already it feels the inborn necessity of finding a home for itself which shall be a rational world a world of permanence and not a kaleidoscopic Wonderland. Point-events, their intervals, the property of $g_{\mu\nu}$ it can make nothing of; these have not the properties it needs. It seeks further, and comes to the quality which we have identified with matter. Here at last is suitable material. Only by developing senses and an imagination which makes this the most real external object can mind find for itself a suitable habitation.[5] The choice is made, and from a fleeting disorder of points and intervals the heavens and the solid earth stand clear.

It must be recognised that the conservation of mass is not exactly equivalent to the permanence of matter. Mind, whilst insisting on a general element of permanence in the things around it would

---

[5]There are other still more complex qualities which would be suitable. If by any chance the mind has preferred one of these, the only difference is a law of gravitation more complicated than that of Einstein, but probably indistinguishable from it experimentally.

have been satisfied with something much less perfect than the actual conservation of mass. The trees put forth leaves, the pond dries up and disappears; for the primitive mind these are definite exceptions, and the fact that delicate measurement traces a conservation of mass even in these cases is scarcely relevant. From another aspect the permanence of matter involves something more than the permanence of mass. When Alice's croquet-mallet turned into a flamingo, it is not necessary to suppose that the conservation of mass was outraged; but a rational mind requires that such incidents should be at any rate uncommon. The continued existence of solid bodies involves laws of nature which are as yet imperfectly understood, and we must leave this difficulty unanswered. Whilst we have not shown that $G_{\mu\nu} - \frac{1}{2}g_{\mu\nu}\,G$ possesses all the qualities desirable for the substance of a perceptual universe, we have shown that it possesses one of the most essential qualities, entirely lacking in any simpler combination; and it is reasonable to think that this had a great deal to do with its selection.

This view of the conditions determining the selection of matter, is strengthened by the consideration that matter does not play such a fundamental part in the analytical world as it does in the perceptual world. The recent tendency of physics has been to regard the quantity known as Action (energy integrated through time) as the most real thing in nature – to put the conclusion crudely. If the perceptual universe were constructed solely in accordance with physical considerations we should expect its substance to be Action. This lack of correspondence has often seemed perplexing, but we can now see that there is good reason for it.

The intervention of mind in the laws of nature is, I believe, more far-reaching than is usually supposed by physicists. I am almost inclined to attribute the whole responsibility for the laws of mechanics and gravitation to the mind, and deny the external world any share in them. It will probably be objected that this is going too far; no doubt the laws depend on the choice made by mind of the material for its universe, but surely Nature deserves some credit for furnishing material with such convenient properties? I doubt it. So far as I can see, all that Nature was required to furnish is a four-dimensional aggregate of point-events; and since these and their relations are undefined, and may be of any character whatever, it should in any case be possible to pick out a set of entities which would serve as point-events, however badly Nature had managed things in the external world. For the use made of the point-events mind alone is responsible.

We have seen that our identification of matter carries with it the laws of conservation of mass, energy, and momentum, and the law of gravitation – in fact, all the laws of mechanics; and further the permanence of matter requires the time and space of experience with all the laws of geometry which belong to the latter. One important group of phenomena remains outside our scheme, viz., the phenomena of electricity, magnetism, and light. A remarkable extension of Einstein's theory has been published recently by Weyl. In this the electromagnetic phenomena find a natural place in the analytical theory. The point of departure from the simpler theory hitherto followed is in the character of the relation called the interval; we have supposed that it is quantitative, so that two distant intervals AB and CD can be immediately compared. Weyl's theory does not admit this comparison at a distance; practically he considers only triangular relations between three neighbouring point-events. It is, of course, impossible to develop the consequences of this without mathematics; but it leads to qualities of the World which can be identified with electromagnetic force, electric charge[6] and current and these automatically satisfy the accepted laws of electromagnetic theory.

If we accept this extension of the theory, it looks at first sight as though all the so-called laws of nature are mere identifications – that the mind singles out for recognition those qualities which as a matter of mathematical identity must necessarily obey the laws it despotically imposes. The laws of mechanics, of electro-dynamics, and of gravitation cover almost the whole field of physics; and yet we have seen that not one of these imposes any constraint on the free arrangement of the external World. Are there then no genuine laws of the external World? Is the universe built from elements which are purely chaotic?

It can scarcely be doubted that our answer must be negative. There *are* laws in the external World, and of these one of the most important (perhaps the only law) is a law of atomicity. We have learnt that a certain quality of the World distinguishes matter from emptiness; we have not learnt why the quality called matter exists only in certain lumps, called atoms or electrons, all of comparable mass. It might be suggested that atomicity arises from a discontinuity in our perceptions which can only vary by finite jumps; but atomicity is not primarily a matter of perception, and the atoms are needed in the analytical theory to account for phenomena which appear continuous to perception. A more likely suggestion is that our analysis into point-events is not final; and if we would carry the analysis beyond the point-event to something still more fundamental, then atomicity and the remaining laws of physics would become obvious identities.

---

[6]The relativity-theory seems almost to ignore the electrical theory of matter which is now so generally accepted; and it even has to contradict the *unqualified* statement that all mass is caused by an electromagnetic field. But there is no real disagreement. The electrical theory of matter has to admit that there is something of unknown nature which holds together the charge of an election; and this extra element in the constitution of matter cannot be ignored in the theory of the gravitational relations of matter.

This may well be the case; and indeed the general attitude of physicists towards theories of nature is that an explanation of this kind is the only one which could be recognised as an ultimate explanation. But the proposed further analysis starts on a different footing from that which we have hitherto conducted. The difference may perhaps be expressed by saying that atomicity specialises the external world, whereas the other laws of physics specialise the mind. I mean that, starting from the postulate that the mind can appreciate only relations, the theory we have described is, or is intended to be, the most general possible theory of the way in which relations can combine to form permanent substance; and accordingly the laws of physics which result depend solely on this postulate as to the mind. Whatever the constitution of the external world,[7] we can pick out a four-dimensional aggregate of entities which we may take to be our point-events since these have been left undefined. But if we attempt to push the analysis behind the point-events, we are, I think, bound to particularise the structure. The investigation, therefore, will begin to distinguish the actual order of nature from other conceivable conditions, and the resulting properties are the true laws of nature.

Whilst we recognise that probably there are true laws of nature, it is perhaps significant that we have not been able to formulate any of them in a general way. Atomicity is manifested not merely in matter, but in connexion with radiation in a large number of phenomena known as quantum phenomena. Our present attitude before these discoveries is one of bewilderment; they have baffled attempts to formulate a general law; and the most successful partial explanations proceed on lines which outrage the canons of thought of the older school of physicists. Thus the domain, where the mind of the physicist has hitherto triumphed, comprises only those laws which have not their seat in the external world, but spring ultimately from the mind. Will the human mind prove equal to formulating the genuine laws of a possibly irrational world, which it has had no part in shaping?

It must be admitted that the atomicity of matter presents a great difficulty from our present point of view. Matter is a property of the world to which the human mind attributes an exaggerated importance for reasons which Nature would regard as irrelevant; yet she seems to be in collusion with mind in singling out this property for atomicity. I can only suggest that the difficulty might disappear if we understood better the true relation between atomicity of matter and the more general atomicity which underlies all quantum phenomena. As far as we can understand it at present, there is some kind of atomicity of the quantity known in mechanics as *Action*, and

---

[7]A sufficient complexity is, of course, required. It is not necessary that the substratum from which we pick out the point-events should be four-dimensional. The straight lines in three-dimensional space form a four-dimensional aggregate.

this seems to be the fundamental origin of all atomic phenomena. If so, that must be Nature's own idea, for which she is in no way indebted to us. On Weyl's theory, Action is chosen because (to put it crudely) it is the only property of the World that could be atomic. Other properties cannot be measured in absolute terms, so that we could attach no meaning to the statement that each atom contains an equal amount of the property; but Action is a pure number, and one unit of Action is a definite amount everywhere. If then we can account for the apparent atomicity of matter as resulting from the quanta of Action, the difficulty alluded to will disappear; but this is at present a speculation.

The physical theories which form the bases of this argument are still on trial, and I am far from asserting that this philosophy of matter is a necessary consequence of discoveries in physics. It is sufficient that we have found one mode of thought tending towards the view that matter is a property of the world singled out by mind on account of its permanence, as the eye ranging over the ocean singles out the waveform for its permanence among the moving waters; that the so-called laws of nature which have been definitely formulated by physicists are implicitly contained in this identification, and are therefore indirectly imposed by the mind; whereas the laws which we have hitherto been unable to fit into a rational scheme, are the true natural laws inherent in the external world, and mind has had no chance of moulding them in accordance with its own outlook.

THE END OF THE WORLD: FROM THE STANDPOINT OF MATHEMATICAL PHYSICS

A. S. Eddington, The End of the World: from the Standpoint of Mathematical Physics, *Nature* **127**, 447-453 (21 March 1931). Presidential address to the Mathematical Association delivered on 5 January 1931.

The world – or space-time – is a four-dimensional continuum, and consequently offers a choice of a great many directions in which we might start off to look for an end; and it is by no means easy to describe "from the standpoint of mathematical physics" the direction in which I intend to go. I have therefore to examine at some length the preliminary question, Which end?

SPHERICAL SPACE

We no longer look for an end to the world in its space dimensions. We have reason to believe that so far as its space dimensions are concerned the world is of spherical type. If we proceed in any direction in space we do not come to an end of space, nor do we continue on to infinity; but, after travelling a certain distance (not inconceivably great), we find ourselves back at our starting point, having 'gone round the world.' A continuum with this property is said to be finite but unbounded. The surface of a sphere is an example of a finite but unbounded two-dimensional continuum; our actual three-dimensional space is believed to have the same kind of connectivity, but naturally the extra dimension makes it more difficult to picture. If we attempt to picture spherical space we have to keep in mind that it is the *surface* of the sphere that is the analogue of our three-dimensional space; the inside and the outside of the sphere are fictitious elements in the picture which have no analogue in the actual world.

We have recently learnt, mainly through the work of Prof. Lemaître, that this spherical space is expanding rather rapidly. In fact if we wish to travel round the world and get back to our starting point, we shall have to move faster than light; because, whilst we are loitering on the way, the track ahead of us is lengthening. It is like trying to run a race in which the finishing-tape is moving ahead faster than the runners. We can picture the stars and galaxies as embedded in the surface of a rubber balloon which is being steadily inflated; so that, apart from their individual motions and the effects of their ordinary gravitational attraction on one another, celestial objects are becoming further and further apart simply by the inflation. It is probable that the spiral nebulae are so distant that they are very little affected by mutual gravitation and exhibit the inflation effect in its pure form. It has been known for some years that they are scattering apart rather rapidly, and we accept their measured rate of recession as a determination of the rate of expansion of the world.

From the astronomical data it appears that the original radius of space was 1200 million light years. Remembering that distances of celestial objects up to several million light years have actually been measured, that does not seem overwhelmingly great. At that radius the mutual attraction of the matter in the world was just sufficient to hold it together and check the tendency to expand. But this equilibrium was unstable. An expansion began, slow at first, but the more widely the matter was scattered the less able was the mutual gravitation to check the expansion. We do not know the radius of space to-day, but I should estimate that it is not less than 10 times the original radius.

At present our numerical results depend on astronomical observations of the speed of scattering apart of the spiral nebulae. But I believe that theory is well on the way to obtaining the same results independently of astronomical observation. Out of the recession of the spiral nebulae we can determine not only the original radius of the universe, but also the total mass of the universe and hence the total number of protons in the world. I find this number to be either 7×10^{78} or 14×10^{78}.[1] I believe that this number is very closely connected with the ratio of the electrostatic and the gravitational units of force, and, apart from a numerical coefficient, is equal to the square of the ratio. If F is the ratio of the electrical attraction between a proton and electron to their gravitational attraction, we find $F^2 = 5.3 \times 10^{78}$. There are theoretical reasons for believing that the total number of particles in the world is αF^2, where α is a simple geometrical factor (perhaps involving π). It ought to be possible before long to find a theoretical value of α, and so make a complete connection between the observed rate of expansion of the universe and the ratio of electrical and gravitational forces.

SIGNPOSTS FOR TIME

---

[1]This ambiguity is inseparable from the operation of counting the number of particles in finite but unbounded space. It is impossible to tell whether the protons have been counted once or twice over.

I must not dally over space any longer but must turn to time. The world is closed in its space dimensions but is open in both directions in its time dimension. Proceeding from 'here' in any direction in space we ultimately come back to 'here'; but proceeding from 'now' towards the future or the past we shall never come across 'now' again. There is no bending round of time to bring us back to the moment we started from. In mathematics this difference is provided for by the symbol $\sqrt{-1}$, just as the same symbol crops up in distinguishing a closed ellipse and an open hyperbola.

If, then, we are looking for an end of the world – or, instead of an end, an indefinite continuation for ever and ever – we must start off in one of the two time directions. How shall we decide which of these two directions to take? It is an important question. Imagine yourself in some unfamiliar part of space-time so as not to be biased by conventional landmarks or traditional standards of reference. There ought to be a signpost with one arm marked 'To the future' and the other arm marked 'To the past.' My first business is to find this signpost, for if I make a mistake and go the wrong way I shall lead you to what is no doubt an 'end of the world,' but it will be that end which is more usually described as the *beginning*.

In ordinary life the signpost is provided by consciousness. Or perhaps it would be truer to say that consciousness does not bother about signposts; but wherever it goes off on urgent business in a particular direction, and the physicist meekly accepts its lead and labels the course it takes 'To the future'. It is an important question whether consciousness in selecting its direction is guided by anything in the physical world. If it is guided, we ought to be able to find directly what it is in the physical world which makes it a one-way street for conscious beings? The view is sometimes held that the 'going on of time' does not exist in the physical world at all and is a purely subjective impression. According to that view the difference between past and future in the material universe has no more significance than the difference between right and left. The fact that experience presents space-time as a cinematograph film which is always unrolled in a particular direction is not a property or peculiarity of the film (that is the physical world) but of the way it is inserted into the cinematograph (that is consciousness). In fact the one-way traffic in time arises from the way our material bodies are geared on to our consciousness:

"Nature has made our gears in such a way That we can never get into reverse."

If this view is right, 'the going on of time' should be dropped out of our picture of the physical universe. Just as we have dropped the old geocentric outlook and other idiosyncrasies of our circumstances as observers, so we must drop the dynamic presentation of events which is no part of the universe itself but is introduced in our peculiar mode of apprehending it. In particular we must be careful not to treat a past-to-future presentation of events as truer or more significant than a future-to-past presentation. We must, of course, drop the theory of evolution, or at least set alongside it a theory of anti-evolution as equally significant.

If anyone here holds this view, I have no argument to bring against him. I can only say to him, "You are a teacher whose duty it is to inculcate in youthful minds a true and balanced outlook. But you teach (or without protest allow your colleagues to teach) the utterly one-sided doctrine of evolution. You teach it not as a colourless schedule of facts but as though there were something significant, perhaps even morally inspiring, in the progress from formless chaos to perfected adaptation. This is dishonest; you should also treat it from the equally significant point of view of anti-evolution and discourse on the progress from future to past. Show how from the diverse forms of life existing to-day Nature anti-evolved forms which were more and more unfitted to survive until she reached the sublime crudity of the paleozoic forms. Show how from the solar system Nature anti-evolved a chaotic nebula. Show how in the course of progress from future to past Nature took a universe which with all its faults is not such a bad effort of architecture and – in short, made a hash of it.

ENTROPY AND DISORGANISATION

Leaving aside the guidance of consciousness, we have found it possible to discover a kind of signpost for time in the physical world. The signpost is of rather a curious character, and I would scarcely venture to say that the discovery of the signpost amounts to the same thing as the discovery of an objective 'going on of time' in the universe. But at any rate it serves to discriminate past and future, whereas there is no corresponding objective distinction of left and right. The distinction is provided by a certain measurable quantity called entropy. Take an isolated system and measure its entropy S at two instants t_1, and t_2. We want to know whether t_1, is earlier or later than t_2 without employing the intuition of consciousness, which is too disreputable a witness to trust in mathematical physics. The rule is that the instant which corresponds to the greater entropy is the later. In mathematical form

$$dS/dt \text{ is always positive.}$$

This is the famous second law of thermodynamics.

Entropy is a very peculiar conception quite unlike the conceptions ordinarily employed in the classical scheme of physics. We may most conveniently describe it as the measure of disorganisation of a system. Accordingly our signpost for time resolves itself into the law that disorganisation increases from past to future. It is one of the most curious features of the development of physics that the entropy outlook grew up quietly alongside the ordinary analytical outlook for a great many years. Until recently it always 'played second fiddle;' it was convenient for getting practical results, but it did not pretend to convey the most penetrating insight. But now it is making a bid for supremacy, and I think there is little doubt that it will ultimately drive out its rival.

There are some important points to emphasise. First there is no other independent signpost for time; so that if we discredit or 'explain away' this property of entropy, the distinction of past and future in the physical world will disappear altogether. Secondly, the test works consistently; isolated systems in different parts of the universe agree in giving the same direction of time. Thirdly, in applying the test we must make certain that our system is strictly isolated. Evolution teaches us that more and more highly organised systems develope as time goes on; but this does not contradict the conclusion that on the whole there is a loss of organisation. It is partly a question of definition of organisation; from the evolutionary point of view it is quality rather than quantity of organisation that is noticed. But,in any case, the high organisation of these systems is obtained by draining organisation from other systems with which they come in contact. A human being as he grows from past to future becomes more and more highly organised – at least he fondly imagines so. But if we make an isolated system of him, that is to say if we cut off his supply of food and drink and air he speedily attains a state which everyone would recognise as 'a state of disorganisation'.

It is possible for the disorganisation of a system to become complete. The state then reached is called thermodynamic equilibrium. The entropy can increase no further, and, since the second law of thermodynamics forbids a decrease, it remains constant. Our signpost for time disappears; and so far as that system is concerned time ceases to go on. That does not mean that time ceases to exist; it exists and extends just as space exists and extends, but there is no longer any one-way property. It is like a one-way street on which there is never any traffic.

Let us return to our signpost. Ahead there is ever-increasing disorganisation. Although the sum total of organisation is diminishing, certain parts of the universe are exhibiting a more and more highly specialised organisation; that is the phenomenon of evolution. But ultimately this must be swallowed up in the advancing tide of chance and chaos, and the whole universe will reach a state of complete disorganisation – a uniform featureless mass in thermodynamic equilibrium. This is the end of the world. Time will *extend* on and on, presumably to infinity. But there will be no definable sense in which it can be said to go on. Consciousness will obviously have disappeared from the physical world before thermodynamical equilibrium is reached, and dS/dt having vanished, there will remain nothing to point out a direction in time.

THE BEGINNING OF TIME

It is more interesting to look in the opposite direction – towards the past. Following time backwards, we find more and more organisation in the world. If we are not stopped earlier, we must come to a time when the matter and energy of the world had the maximum possible organisation. To go back further is impossible. We have come to an abrupt end of space-time – only we generally call it the 'beginning'.

I have 'no philosophical axe to grind' in this discussion. Philosophically, the notion of a beginning of the present order of Nature is repugnant to me. I am simply stating the dilemma to which our present fundamental conception of physical law leads us. I see no way round it; but whether future developments of science will find an escape I cannot predict. The dilemma is this: Surveying our surroundings we find them to be far from a 'fortuitous concourse of atoms.' The picture of the world, as drawn in existing physical theories, shows arrangement of the individual elements for which the odds are multillions[2] to 1 against an origin by chance. Some people would like to call this non-random feature of the world purpose or design; but I will call it non-committally anti-chance. We are unwilling to admit in physics that anti-chance plays any part in the reactions between the systems of billions of atoms and quanta that we study; and indeed all our experimental evidence goes to show that these are governed by the laws of chance. Accordingly, we sweep anti-chance out of the laws of physics – out of the differential equations. Naturally, therefore, it reappears in the boundary conditions, for it must be got into the scheme somewhere. By sweeping it far enough away from the sphere of our current physical problems, we fancy we have got rid of it. It is only when some of us are so misguided as to try to get back billions of years into the past that we find the sweepings all piled up like a high wall and forming a boundary – a beginning of time-which we cannot climb over.

A way out of the dilemma has been proposed which seems to have found favour with a number of scientists. I oppose it because I think it is untenable, not because of any desire to retain the present dilemma. I should like to find a genuine loophole. But that does not alter my conviction that the loophole that is at present being advocated is a blind alley. I must first deal with a minor criticism.

I have sometimes been taken to task for not sufficiently emphasising in my discussion of these problems that the results about entropy are a matter of probability, not of certainty. I said above that if we

---

[2]I use "multillions" as a general term for numbers of order $10^{10^{10}}$ or larger.

observe a system at two instants the instant corresponding to the greater entropy will be the later. Strictly speaking I ought to have said that for a smallish system the chances are, say, 10^{20} to 1, that it is the later. Some critics seem to have been shocked at my lax morality in making such a statement, when I was well aware of the 1 in 10^{20} chance of its being wrong. Let me make a confession. I have in the past 25 years written a good many papers and books, broadcasting a large number of statements about the physical world. I fear that for not many of these statements is the risk of error so small as 1 in 10^{20}. Except in the domain of pure mathematics the trustworthiness of my conclusions is usually to be rated at nearer 10 to 1 than 10^{20} to 1; even that may be unduly boastful. I do not think it would be for the benefit of the world that no statement should be allowed to be made if there were a 1 in 10^{20} chance of its being untrue. Conversation would languish somewhat. The only persons entitled to open their mouths would presumably be the pure mathematicians.

FLUCTUATIONS

The loophole to which I referred depends on the occurrence of chance fluctuations. If we have a number of particles moving about at random they will in the course of time go through every possible configuration, so that even the most orderly, the most non-chance, configuration will occur by chance if only we wait long enough. When the world has reached complete disorganisation (thermodynamic equilibrium) there is still infinite time ahead of it, and its elements will thus have opportunity to take up every possible configuration again and again. If we wait long enough, a number of atoms will, just by chance, arrange themselves in systems as they are at present arranged in this room; and, just by chance, the same sound-waves will come from one of these systems of atoms as are at present emerging from my lips; they will strike the ears of other systems of atoms, arranged just by chance to resemble you, and in the same stages of attention or somnolence. This mock Mathematical Association meeting must be repeated many times over – an infinite number of times in fact – before t reaches $+\infty$. Do not ask me whether I expect you to believe that this will really happen:[3]

"Logic is logic. That's all I say."

So after the world has reached thermodynamical equilibrium the entropy remains steady at its maximum value, except that 'once in a blue moon' the absurdly small chance comes off and the entropy drops appreciably below its maximum value. When this fluctuation

---

[3] I am hopeful that the doctrine of the "expanding universe" will intervene to prevent its happening.

has died out there will again be a very long wait for another coincidence giving another fluctuation. It will take multillions of years, but we have all infinity of time before us. There is no limit to the amount of the fluctuation, and if we wait long enough we shall come across a big fluctuation which will take the world as far from thermodynamical equilibrium as it is at the present moment. If we wait for an enormously longer time, during which this huge fluctuation is repeated untold numbers of times, there will occur a still larger fluctuation which will take the world as far from thermodynamic equilibriumas it was one second ago.

The suggestion is that we are now on the downward slope of one of these fluctuations. It has quite a pleasant subtlety. Is it chance that we happen to be running down the slope and not toiling up the slope? Not at all. So far as the physical universe is concerned, we have *defined* the direction of time as the direction from greater to less organisation, so that on whichever side of the mountain we stand our signpost will point downhill. In fact, on this theory, the going on of time is not a property of time in general, but is a property of the slope of the fluctuation on which we are standing. Again, although the theory postulates a universe involving an extremely improbable coincidence, it provides an infinite time during which the most improbable coincidence might occur. Nevertheless I feel sure that the argument is fallacious.

If we put a kettle of water on the fire there is a chance that the water will freeze. If mankind goes on putting kettles on the fire until $t = \infty$, the chance will one day come off and the individual concerned will be somewhat surprised to find a lump of ice in his kettle. But it will not happen to *me*. Even if to-morrow the phenomenon occurs before my eyes I shall not explain it this way. I would much sooner believe in the interference by a demon than in a coincidence of that kind coming off; and in doing so I shall be acting as a rational scientist. The reason why I do not at present believe that devils interfere with my cooking arrangements and other business, is because I have become convinced by experience that Nature obeys certain uniformities which we call laws. I am convinced because these laws have been tested over and over again. But it is possible that every single observation from the beginning of science, which has been used as a test, has just happened to fit in with the law by a chance coincidence. It is an improbable coincidence, but I think not quite so improbable as the coincidence involved in my kettle of water freezing. So if the event happens and I can think of no other explanation, I shall have to choose between two highly improbable coincidences: (a) that there are no laws of Nature and that the apparent uniformities so far observed are merely coincidences; (b) that the event is entirely in accordance with the accepted laws of Nature, but that an improbable coincidence has happened. I choose the former because mathematical calculation indicates that it is the less improbable. I reckon a sufficiently improbable coincidence as something much more disastrous than a violation of the laws of Nature; because my whole reason for accepting the laws of Nature rests on the assumption that improbable coincidences do not happen – at least that they do not happen in my experience.[4]

Similarly, if logic predicts that a mock meeting of the Mathematical Association will occur just by a fortuitous arrangement of atoms before $t = \infty$, I reply that I cannot possibly accept that as being the explanation of a meeting of the Mathematical Association in $t = 1931$. We must be a little careful over this, because there is a trap for the unwary. The year 1931 is not an absolutely random date between $t = -\infty$ and $t = +\infty$. We must not argue that because for only $1/x$th of time between $t = -\infty$ and $t = +\infty$ a fluctuation as great as the present one is in operation, therefore the chances are x to 1 against such a fluctuation occurring in the year 1931. For the purposes of the present discussion the important characteristic of the year 1931 is that it belongs to a period during which there exist in the universe beings capable of speculating about the universe and its fluctuations. Now I think it is clear that such creatures could not exist in a universe in thermodynamical equilibrium. A considerable degree of deviation is required to permit of living beings. Therefore it is perfectly fair for supporters of this suggestion to wipe out of account all those multillions of years during which the fluctuations are less than the minimum required to permit of the development and existence of mathematical physicists. That greatly diminishes x, but the odds are still overpowering. The *crude* assertion would be that (unless we admit something which is not chance in the architecture of the universe) it is practically certain that at any assigned date the universe will be almost in the state of maximum disorganisation. The *amended* assertion is that (unless we admit something which is not chance in the architecture of the universe) it is practically certain that a universe containing mathematical physicists will at any assigned date be in the state of maximum disorganisation which is not inconsistent with the existence of such creatures. I think it is quite clear that neither the original nor the amended version applies. We are thus driven to admit anti-chance; and apparently the best thing we can do with it is to sweep it up into a heap at the beginning of time, as I have already described.

The connection between our entropy signpost and that dynamic quality of time which we describe as 'going on' or 'becoming' leads to very difficult questions which I cannot discuss here. The puzzle is that the signpost seems so utterly different from the thing of which

---

[4]No doubt "extremely improbable" coincidences occur to all of us, but the improbabillty is of an utterly different order of magnitude from that concerned in the present discussion.

it is supposed to be the sign. The one thing on which I have to insist is that, apart from consciousness, the increase of entropy is the only trace that we can find of a one-way direction of time. I was once asked a ribald question: How does an electron (which has not the resource of consciousness) remember which way time is going? Why should it not inadvertently turn round and, so to speak, face time the other way? Does it have to calculate which way entropy is increasing in order to keep itself straight? I am inclined to think that an electron does do something of that sort. For an electric charge to face the opposite way in time is the same thing as to change the sign of the charge. So if an electron mistook the way time was going it would turn into a positive charge. Now, it has been one of the troubles of Dr. P. A. M. Dirac that in the mathematical calculations based on his wave equation the electrons do sometimes forget themselves in this way. As he puts it, there is a finite chance of the charge changing sign after an encounter. You must understand that they only do this in the mathematical problems, not in real life. It seems to me there is good reason for this. A mathematical problem deals with, say, four electric charges at the most; that is about as many as a calculator would care to take on. Accordingly, the unfortunate electron in the problem has to make out the direction of past to future by watching the organisation of three other charges. Naturally, it is deceived sometimes by chance coincidences which may easily happen when there are only three particles concerned; and so it has a good chance of facing the wrong way and becoming a positive charge. But in any real experiment we work with apparatus containing billions of particles – ample to give the electron its bearings with certainty. Dirac's theory predicts things which never happen, simply because it is applied to problems which never occur in Nature. When it is applied to four particles alone in the universe, the analysis very properly brings out the fact that in such a system there could be no steady one-way direction of time, and vagaries would occur which are guarded against in our actual universe consisting of about 10^{79} particles.

HEISENBERG'S PRINCIPLE

A discussion of the properties of time would be incomplete without a reference to the principle of indeterminacy, which was formulated by Heisenberg in 1927 and has been generally accepted. It had already been realised that theoretical physics was drifting away from a deterministic basis; Heisenberg's principle delivered the knock-out blow, for it actually postulated a certain measure of indeterminacy or unpredictability of the future as a fundamental law of the universe. This change of view seems to make the progress of time a much more genuine thing than it used to be in classical physics. Each passing moment brings into the world something new – something which is not merely a mathematical extrapolation of what was already there.

The deterministic view which held sway for at least two centuries was that if we had complete data as to the state of the whole universe during, say, the first minute of the year 1600, it would be merely a mathematical exercise to deduce everything that has happened or will happen at any date in the future or past. The future would be determined by the present as the solution of a differential equation is determined by the boundary conditions. To understand the new view it is necessary to realise that there is a risk of begging the question when we use the phrase 'complete data.' All our knowledge of the physical world is inferential. I have no direct acquaintance with my pen as an object in the physical world; I infer its existence and properties from the light waves which fall on my eyes, the pressure waves which travel up my muscles, and so on. And precisely the same scheme of inference leads us to infer the existence of things in the past. Just as I infer a physical object, namely my pen, as the cause of certain visual sensations now, so I may infer an infection some days ago as the cause of an attack of measles. If we follow out this principle completely we shall infer causes in the year 1600 for all the events which we know to have happened in 1930. And at first sight it would seem that these inferred causes have just as much status in the physical world as my fountain pen, which is likewise an inferred cause. So the determinist thinks he has me in a cleft stick. If the scientific worker poking about in the universe in 1600 comes across these causes, then he has all the data for making a correct prediction for 1930; if he does not, then he clearly has not complete knowledge of the universe in 1600, for these causes have as much right to the status of physical entities as any of our other inferences.

I need scarcely stop to show how this begs the question by arbitrarily prescribing what we should deem to be complete knowledge of the universe in 1600, irrespective of whether there is any conceivable way in which this knowledge could be obtained at the time. What Heisenberg discovered was that (at least in a wide range of phenomena embracing the whole of atomic physics and electron theory) there is a provision of Nature that just half of the data demanded by our determinist friend might with sufficient diligence be collected by the investigators in 1600, and that complete knowledge of this half would automatically exclude all knowledge of the other half. It is an odd arrangement, because you can take your choice which half you will find out; you can know either half but not both halves. Or you can make a compromise and know both halves imperfectly, that is with some margin of uncertainty. But the rule is definite. The data are linked in pairs and the more accurately you measure one member of the pair the less accurately you can measure the other member.

Both halves are necessary for a complete prediction of the future, although, of course, by judiciously choosing the type of event we predict we can often make safe prophecies. For example, the principle of indeterminacy will obviously not interfere with my prediction that during the coming year zero will turn up approximately $\frac{1}{37}$ of the total number of times the roulette ball is spun at Monte Carlo. All our successful predictions in physics and astronomy are on examination found to depend on this device of eliminating the inherent uncertainty of the future by averaging.

As an illustration, let us consider the simplest type of prediction. Suppose we have a particle, say an electron, moving undisturbed with uniform velocity. If we know its position now and its velocity, it is a simple matter to predict its position at some particular future instant. Heisenberg's principle asserts that the position and velocity are paired data; that is to say, although there is no limit to the accuracy with which we might get to know the position and no limit to the accuracy with which we might get to know the velocity, we cannot get to know both. So our attempt at an accurate prediction of the future position of the particle is frustrated. We can, if we like, observe the position now and the position at the future instant with the utmost accuracy (since these are not paired data) and then calculate what has been the velocity in the meantime. Suppose that we use this velocity together with the original position to compute the second position. Our result will be quite correct, and we shall be true profits – after the event.

This principle is so fully incorporated into modern physics that in wave mechanics the electron is actually pictured in a way which exhibits this 'interference' of position and velocity. To attribute to it exact position and velocity simultaneously would be inconsistent with the picture. Thus according to our present outlook the absence of one half of the data of prediction is not to be counted as ignorance; the data are lacking because they do not come into the world until it is too late to make the prediction. They come into existence when the event is accomplished.

I suppose that to justify my title I ought to conclude with a prophecy as to what the end of the world will be like. I confess I am not very keen on the task. I half thought of taking refuge in the excuse that, having just explained the future is unpredictable, I ought not to be expected to predict it. But I am afraid that someone would point out that the excuse is a thin one, because all that is required is a computation of averages and that type of prediction is not forbidden by the principle of indeterminacy. It used to be thought that in the end all the matter of the universe would collect into one rather dense ball at uniform temperature; but the doctrine of spherical space, and more especially the recent results as to the expansion of the universe have changed that. There are one or two unsettled points which prevent a definite conclusion, so I will content myself with stating one of several possibilities. It is widely thought that matter slowly changes into radiation. If so, it would seem that the universe will ultimately become a ball of radiation growing ever larger, the radiation becoming thinner and passing into longer and longer wavelengths. About every 1500 million years it will double its radius, and its size will go on expanding in this way in geometrical progression for ever.

THE EXPANDING UNIVERSE

A. S. Eddington, The Expanding Universe, *Nature* **129**, 421-423 (19 March 1932). Friday evening discourse delivered at the Royal Institution on 22 January 1932.

In recent years the line-of sight velocities of about ninety of the spiral nebulae have been measured. The distances of some of the nearest of them have been determined by a fairly trustworthy method and for others rude estimates depending on statistical methods are available. When we survey these data, a remarkable state of affairs is revealed. The spiral nebulae are almost unanimously running away from us; moreover, the greater the distance the greater the speed of recession. The law of increase is found to be fairly regular, the speed being simply proportional to the distance. The progression has been traced up to a distance of more than 100 million light-years, where the recession is 20,000 km. per sec. – about the speed of an α–particle.

At first sight this looks as though the spiral nebulae had a rather pointed version to our society; but a little consideration will show that the phenomenon is merely a uniform dilation of the system and is not specially directed at us. If this room were suddenly to expand to twice its present size, the seats separating in proportion, you would notice that everyone in the room had moved away from *you*. Your neighbor who was 3 feet away has become 6 feet away; the man over yonder who was 20 feet away is now 40 feet away. Each has moved proportionately to his distance from you, which is precisely what the spiral nebulae are observed to be doing. The motion is not directed from any one center, but is a general expansion, such that each individual observes every other individual to be receding.

In 1917, before any hint of this phenomena had been obtained from observation, Prof. W. de Sitter was on the look out for something of the kind. He found that, on one of two alternative hypotheses arising out of Einstein's relativity theory, the light of very remote objects should be displaced to the red as though they were moving away from us; and he suggested the observed motions of the spiral nebulae (by far the most remote objects known) as a discriminating test. At that time, only three radial velocities had been published, and they rather lamely supported his hypotheses by a majority of 2 to 1. The majority has now become about 85 to 5, and the five exceptions are nebulae close to us which in any case should have had only small receding velocities. De Sitter's theory has been developed and modified by Friedmann[1] and Lemaître; the modern view of it is as follows:

Einstein' law of gravitation contains a term called the 'cosmical term' which is extremely small in ordinary applications to the solar system, etc., and is generally neglected. The term, however, actually represents a repulsive force directly proportional to the distance; so that however small it may be in ordinary applications, if we go to distances sufficiently great it must ultimately become important. It is this cosmical repulsion which is, we believe, the cause of the expansion of the great system of the nebulae. The repulsion may be to some extent counterbalanced by the ordinary gravitational attraction of the nebulae on one another. This countervailing attraction will become weaker as the expansion increases and the nebulae become further apart. It seems likely that the universe started with a balance between gravitational attraction and cosmical repulsion; this equilibrium states is called an 'Einstein universe'. But it can be shown that the Einstein universe is unstable; and the slightest disturbance will cause either the repulsion or the gravitation to gain the upper hand, so as to topple the system into a state of continually increasing expansion or continually increasing contraction. Apparently, expansion won the initial struggle, and as the nebulae spread apart, the opposition of gravitation became less and less, until now it is comparatively insignificant.

We see, then, that according to observation the system of the spiral nebulae is expanding, and that relativity theory had foreseen just an expansion (except that as an alternative it would have been content with an equally regular contraction). What better agreement could we desire? Nevertheless, there were some misgivings which I would not by any means condemn as unreasonable. It is true that theory predicted an effect of the kind observed, but it did not say how rapid the expansion would be. It expressed it in terms of an unknown 'cosmical constant' λ, leaving λ to be determined by observation. Now the rate of expansion indicated by observation comes to us as a great shock. The universe is expanding so as to double its dimensions every 1300 million years; that is no more than the period of geological time. Astronomers, who had been picturing a slow evolution of the stars extending over billions of years, would, scarcely believe our staid old universe capable of such a hustle. In fact, it

---

[1] EDITOR'S NOTE: A. A. Friedmann, *Papers On Curved Spaces and Cosmology* (Minkowski Institute Press, Montreal 2014) and (for a wider audience) A. A. Friedmann, *The World as Space and Time* (Minkowski Institute Press, Montreal 2014).

means a cut of something like ninety-nine per cent in our time-scale, which even in these days of economy cuts is not to be accepted lightly by the department concerned. For this reason many have thought that the receding motions of the spiral nebulae cannot be accepted as genuine, and that the whole phenomenon must be explained away as a misinterpretation of the red-shift observed in their spectra.

I think, however, that we shall have to accept the expansion. My reason is that it now seems possible to calculate the cosmical constant λ by pure physical theory. The value is the same as that given by the recession of the nebulae, so that there is full confirmation.

I have been tracing the effects of the cosmical constant in the behaviour of the great system of galaxies – phenomena on the grandest scale we have yet imagined. Now I want to turn to the other end of the scale and look into the interior of the atom, where, I think, we shall find that the same cosmical constant turns up again. It is, in fact, the main key to the mystery of protons and electrons. I cannot go very far into this part of the theory, but I will try to show why I am convinced that the cosmical constant comes into the theory of the atom. I must premise one thing. It is well known that, in Einstein's theory, gravitation has an interpretation not only as a force but also as a geometrical property – a curvature – of space-time. So also the cosmical constant has an interpretation not only in terms of repulsive force but also as a measure of curvature. The constant λ is, in fact, precisely equal to $1/R^2$, R being the radius of the world in the equilibrium (Einstein) state from which we suppose it to have started.

Length is necessary relative. That is one of the results of Einstein's theory which has become almost a commonplace of physics; but it was a rather complicated kind of relativity that Einstein considered – relativity to the motion of our frame of reference. I am going to refer to another much more elementary relativity of length, namely, that length always implies comparison with a standard of length. It is only the ratio of length that enters into our experience. Suppose that every length and every distance in the universe were suddenly to be doubled; nothing would seem altered. I do not think we could attach any meaning to the change. Intrinsically, Brobdingnag and Lilliput are precisely the same; it needs an intruding Gulliver – an extraneous standard of length – to make them appear different.

Now, it is commonly stated in physics that all normal hydrogen atoms have the same size, or have the same spread of electric charge. We have a very fundamental equation (the wave equation) determining the spread, which is supposed to apply to any hydrogen atom and, of course, gives the same result for all. But what do we mean by their having the same size? Or it may be better to put the question negatively – What would it mean if we said that two hydrogen atoms were of different sizes, that is similarly constructed but on different scales? It would be Brobdingnag and Lilliput over again. To give any meaning to the difference, we need a Gulliver. Now, the Gulliver of physics is always supposed to be a certain bar of metal called the International Metre. He is anything but a traveller; I think he has never been away from Paris. It was Prof. Weyl who first directed attention to the very big hiatus involved, when we speak of a length such as the radius of hydrogen atom being a certain fraction of a standard metre. We have, as it were, our Gulliver but have left out his travels. The travels are (as Weyl showed) the interesting part of the story, and are not to be glossed over as irrelevant.

Weyl went further and pointed out that there is a natural standard of comparison which is always on the spot, namely, the radius of curvature of the world *at that spot*. We can thus give a direct meaning to the statement that two hydrogen atoms in any part of the universe have the same size; we mean that each of them is the same fraction of the radius of curvature of space-time at the place where it lies. The atom here is a particular fraction of the radius of curvature here; the atom on Sirius is the same fraction of the radius of curvature at Sirius. Whether the radius of curvature here is the same as at Sirius does not arise, and I do not think there is any meaning in trying to compare them.

The above definition of equality, and the use of Weyl's standard, may seem a dangerous innovation; but, indirectly, we have been using it all along, without knowing that we were doing so. Some years ago I pointed out that Einstein's law of gravitation can be stated in the form, "What we call a metre at any place and in any direction is a constant fraction of the radius of curvature of space-time for that place and direction". That is simply a translation of the law from symbols into words. The law is verified by observation, so that the statement gives us not only an ideal definition of the metre but also one which we know will accord with the reckoning of metres that is actually used. Thus, measurement in terms of the metre is equivalent to measurement in terms of the world radius, since the two standards are always in a constant ratio. Practically, it is more convenient to employ the metre, but in pursuing the theory we must go direct to the world radius; for obviously a particular bar of metal at Paris can have no fundamental status in physics and is altogether irrelevant to equations describing the mechanism of the atom. The world curvature, on the other hand, is on the spot and is directly reacting with the atom.

I now return to the wave equation which professes to determine how large an atom will be. That, as we have seen, means that it finds the ratio of the various intervals in the atom to the world radius there; so the world radius must come into the equation. But the world radius is the cosmical constant in another form. The cosmical constant has cropped up again inside the atom.

My task now was to spot the cosmical constant or the world radius in the current form of the wave equation, which is known by experiment to be substantially correct. It is very much disguised, because the current equation introduces the standard metre and all sorts of irrelevancies. But one knows the sort of effect that curvature can have; and the way it will appear in the equation is pretty well dictated by the quantum laws, which make a speciality of the properties of 'closed circuits' such as are introduced by curved space. I think I succeeded, and I arrived at the identification

$$\frac{mc^2}{e^2} = \sqrt{\frac{N}{R}}.$$

The left is a term in the current wave equation, and its value is known experimentally. The right side is the way that we write it now that we have penetrated its disguise. R is the Einstein radius of the world, equal to the universe square root of the cosmical constant; N is the number of electrons (or protons) in the universe.

This additional equation, combined with other equations already known, gives all the information required. We deduce, for example, that the number of electrons in the universe is 1.29×10^{79}; and that the original radius of the universe, before it started to expand, was 1070 million light-years. Most important of all, we find that the consequent rate of expansion of the universe is 528 km. per sec. per megaparsec distance. The observational determinations from the recession of the spiral nebulae (which might be a little lower, since they include any countervailing gravitational attraction) range from 430 km. to 55 km. per sec. per megaparsec. We can feel little doubt, therefore, that the observed motions of the nebulae are genuine and represent the expansion effect predicted by relativity. We must reconcile ourselves to this alarming rate of expansion, which plays havoc with older ideas as to the time-scale.

However interesting may be the application of this theory to the universe, the application to the interior of the atom seems likely to be still more fruitful. Now that we know the magnitude of the radius of curvature, we can set aside the arbitrary metre and use this natural unit in our equations. The big uninformative coefficient disappear; and the equations are so much simplified that, I think, I have a fair idea of what they really mean and how they work. In particular, the relation of the proton to the electron is now apparent and the theoretical ratio of their masses is found to be 1847.6; this is certainly very near to the observed value.

I do not want to stress too much the accuracy or finality of these first results. I cannot see how anything can possible be wrong with them; but then one never does see these faults until some new circumstance arises or some ingenious person comes forward to show us how blind we have been. At least, a way of progress has been found. I think that some day, when electrons and protons have come to order, we shall look back and see that the key to the mystery was lying somewhere in intergalactic space and was picked up by astronomers who measured the velocities and distances of nebulae ten million light-years away.

THE DECLINE OF DETERMINISM

A. S. Eddington, The Decline of Determinism, *Nature* **129**, 233-240 (13 February 1932). Presidential address to the Mathematical Association, delivered on 4 January 1932.

DETERMINISM has faded out of theoretical physics. Its exit has been commented on in various ways. Some writers are incredulous, and cannot be persuaded that determinism has really been eliminated. Some think that it is only a domestic change in physics having no reactions on general philosophical thought. Some imagine that it is a justification for miracles. Some decide cynically to wait and see if determinism fades in again.

The rejection of determinism is in no sense an abdication of scientific method; indeed it has increased the power and precision of the mathematical analysis of observed phenomena. On the other hand, I cannot agree with those who belittle the general philosophical significance of the change. The withdrawal of physical science from an attitude it has adopted consistently for more than two hundred years is not to be treated lightly; and it involves a reconsideration of our views with regard to one of the perplexing problems of our existence. In this address, I shall deal mainly with the physical universe, and say very little about mental determinism or freewill. That might well be left to those who are more accustomed to arguing about such questions, if only they could be awakened to the new situation which has arisen on the physical side. At present I can see little sign of such an awakening.

DEFINITIONS OF DETERMINISM

Let us first be sure that we agree as to what is meant by determinism. I quote three definitions or descriptions for your consideration. The first is by a mathematician (Laplace):

> We ought then to regard the present state of the universe as the effect of its antecedent state and the cause of the state that is to follow. An intelligence. who for a given instant should be acquainted with all the forces by which Nature is animated and with the several positions of the

entities composing it, if further, his intellect were vast
enough yo submit those data to analysis, would include
in one and the same formula the movements of the largest
bodies in the universe and those of the lightest atom.
Nothing would be uncertain for him; the future as well
as the past would be present to his eyes. The human mind
in the perfection it has been able to give to astronomy
affords a feeble outline of such an intelligence. ... All
its efforts in the search for truth tend to approximate
without limit to the intelligence we have just imagined.

The second is by a philosopher (C.D. Broad):

'Determinism' is the name given to the following doctrine. Let S be any substance, ψ any characteristic, and
t any moment. Suppose that S is in fact the state σ with
respect to ψ at t. Then the compound supposition that
everything else in the world should have been exactly as
it in fact was, and that as should have been in one of
the other two alternative states with respect to ψ is an
impossible one. [The three alternative states (of which σ
is one) are to have the characteristic ψ, not to have it,
and to be changing.]

The third is by a poet (Omar Khayyám):

With Earth's first Clay They did the Last Man knead,
And there of the Last Harvest sow'd the Seed: And the
first Morning of Creation wrote What the Last Dawn of
Reckoning shall read.

I proposed to take the poet's description as my standard. Perhaps this may seem an odd choice; but there is no doubt that his
words express what is in our minds when we refer to determinism.
The other two definitions need to be scrutinized suspiciously; we are
afraid they may be a catch in them. In saying that the physical universe as now pictured is not a universe in which "the first morning
of creation wrote what the last dawn of reckoning shall read", we
make it clear that the abandonment of determinism is no technical
quibble, but is to be understood in the most ordinary sense of the
words.

It is important to notice that all three definitions introduce the
time-element. Determinism postulates not merely causes but pre-
existing causes. Determinism means predetermination. Hence in
any argument about determinism the dating of the alleged causes is
an important matter; we must challenge them to produce their birth
certificates.

Ten years ago, practically every physicist of repute was, or believed himself to be, a determinist, at any rate so far as inorganic phenomena are concerned. He believed that he had come across a scheme of strictly causal law, and that it was the primary aim of science to fit as much of our experience as possible into such a scheme. The methods, definitions, and conceptions of physical science were so much bound up with this assumption of determinism that the limits (if any) of the scheme of casual law were looked upon as the ultimate limits of physical science.

To see the change that has occurred, we need only refer to a recent book which goes as deeply as anyone has yet penetrated into the fundamental structure of the physical universe, Dirac's "Quantum Mechanics". I do not know whether Dirac is a determinist or not; quite possibly he believes as firmly as ever in the existence of a scheme of strict causal law. But the significant thing is that in this book he has no occasion to refer to it. In the fullest account of what has yet been ascertained as to the way things work, casual law is not mentioned.

This is a deliberate change in the aim of theoretical physics. If the older physicist had been asked why he thought that progress consisted in fitting more and more phenomena into a deterministic scheme, his most effective reply would have been "What else is there to do?" A book such as Dirac's supplies the answer. For the new aim has been extraordinarily fruitful, and phenomena which had hitherto baffled exact mathematical treatment are now calculated and the predictions are verified by experiment. We shall see presently that indeterministic law is as useful a basis for practical predictions as deterministic law was. By all practical tests, progress along this new branch track must be recognised as a great advance in knowledge. No doubt some will say "Yes, but it is often necessary to make a detour in order to get round an obstacle. Presently we shall have past the obstacle and be able to join the old road again." I should say rather that we are like explorers on whom at last it has dawned that there are other enterprizes worth pursuing besides finding the North-West Passage; and we need not take too seriously the prophesy of the old mariners who regard these enterprises as a temporary diversion to be followed by a return to the 'true aim of geographical exploration'. But at the moment I am concerned with prophecy ans counter-prophecy; the important thing is to grasp the facts of the present situation.

Let us first try to see how the new aim of physical science originated. We observe certain regularities in the course of Nature and formulate these as 'laws of Nature'. Laws may be stated positively or negatively, 'Thou shalt' or 'Thou shalt not'. For the present purpose it is most convenient to formulate them negatively. Consider the following two regularities which occur in our experience:

(a) We never come across equilateral triangles whose angles are unequal.

(b) We never come across 13 trumps in our hand at bridge.

In our ordinary outlook we explain these regularities in fundamentally different ways. We say that the first occurs because the contrary experience *is impossible*; the second occurs because the contrary experience is *too improbable*.

This distinction is entirely theoretical; there is nothing in the observations themselves to suggest to which type a particular regularity belongs. We recognise that 'impossible' and 'too improbable' can both give adequate explanation of any observed uniformity of experience, and the older theory rather haphazardly explained some uniformities one way and other uniformities the other way. In the new physics we make no such discrimination; the union obviously must be on the basis of (b), not (a). It can scarcely be supposed that there is a law of Nature which makes the holding of 13 trumps in a properly dealt hand impossible; but it *can* be supposed that our failure to find equilateral triangles with unequal angles is only because such triangles are too improbable.

We must, however, first consider the older view which distinguished type (a) as a special class of regularity. Accordingly, there were two types of natural law. The earth keeps revolving round the sun, because it is *impossible* it should run away. Heat flows from a hot body to a cold, because it is *too improbable* that it should flow the other way. I call the first type *primary* law, and the second type *secondary* law. The recognition of secondary law was the thin end of the wedge that ultimately cleft the deterministic scheme.

For practical purposes primary and secondary law exert equally strict control. The improbability referred to in secondary law is so enormous that failure even in an isolated case is not to be seriously contemplated. You would be utterly astounded if heat flowed from you to the fire so that you got chilled by standing in front of it, although such an occurrence is judged by physical theory to be not impossible but improbable. Now it is axiomatic that in a deterministic scheme nothing is left to chance; a law which has to ghost of a chance of failure cannot form part of the scheme. So long as the aim of physics is to bring to light a deterministic scheme, the pursuit of secondary law is a blind alley since it leads only to probabilities. The determinist is not content with a law which prescribes that, given reasonable luck, the fire will warm me; he admits that that is the probable effects, but adds that somewhere at the base of physics there are other laws which prescribe just what the fire will do to me, luck or no luck.

To borrow an analogy from genetics, determinism is a *dominant character*. We can (and indeed must) have secondary indeterministic laws within any scheme of primary deterministic law – laws which tell us what is likely to happen but are overridden by the dominant laws which tell us what must happen. So determinism watched with equanimity the development of indeterministic law within itself. What matter? Deterministic law remains dominant. It was not foreseen that indeterministic law when fully grown might be able to stand by itself and supplant its dominant parent. There is a game called "Think of a number". After doubling, adding, and other calculations, there comes the directions "Take away the number you first thought of". We have reached that position in physics, and the time has come to take away the determinism we first thought of.

The growth of secondary law within the deterministic scheme was remarkable and gradually sections of the subject formerly dealt with by primary law were transferred to it. There came a time when in some of the most progressive branches of physics secondary law was used exclusively. The physicist might continue to profess allegiance to primary law but he ceased to utilize it. Primary law was the gold to be kept stored in vaults; secondary law was the paper to be used for actual transactions. No one minded; it was taken for granted that the paper was backed by gold. At last came the crisis, and *physics went off the gold standard.* This happened very recently, and opinions are divided as to what the result will be. Prof. Einstein, I believe, fears disastrous inflation, and urges a return to sound currency – if we can discover it. But most theoretical physicists have begun to wonder why the now idle gold should have been credited with such magic properties. At any rate the thing has happened, and the immediate result has been a big advance in atomic physics.

We have seen that indeterministic or secondary accounts for regularities of experience, so that it can be used for predicting the future as satisfactorily as primary law. The predictions and regularities refer to average behaviour of the vast number of particles concerned in most of our observations. When we deal with fewer particles the indeterminacy begins to be appreciable, and prediction becomes more of a gamble; until finally the behaviour of a single atom or electron has a very large measure of indeterminacy. Although some courses may be more probable than others, backing an electron to do anything is in general as uncertain as backing a horse.

It is commonly objected that our uncertainty as to what the electron will do in the future is due not to indeterminism but to ignorance. It is asserted that some character exists in the electron or its surroundings which decides its future, only physicists have not yet learned how to detect it. You will see later how I deal with this suggestions. But I would here point out that if the physicist is to take any part in the wider discussion on determinism as affecting the significance of our lives and the responsibility of our decisions, he must do so on the basis of what he has discovered, not on the basis of what it is conjectured he might discover. His first step should be to make clear that he no longer holds the position, occupied for so long, of chief advocate for determinism, and that he is *unaware* of any deterministic law in the physical universe. He steps aside and leaves it to others – philosophers, psychologists, theologians – to come forward and show, if they can, that they have found indications of determinism in some other way.[1] If no one comes forward, the hypothesis of determinism presumably drops; and the question whether physics is actually antagonistic to it scarcely arises. It is no use looking for an opposer until there is a proposer in the field.

INFERENTIAL KNOWLEDGE

It is now necessary to examine rather closely the nature of our knowledge of the physical universe.

All our knowledge of physical objects is by inference. We have no means of getting into direct contact with them; but they emit and scatter light waves, and they are the source of pressures transmitted through adjacent material. They are like broadcasting stations that send out signals which we can receive. At one stage of the transmission the signals pass along nerves within our bodies. Ultimately visual, tactical, and other sensations are provoked in the mind. It is from these remote effects that we have to argue back to the properties of the physical object at the far end of the chain of transmission. The image which arises in the mind is not the physical object, though it is a source of information about the physical object; to confuse the mental object with the physical object is to confuse the clue with the criminal. Life would be impossible if there were no kind of correspondence between the external world and the picture of it in our minds; and natural selection (reinforced where necessary by the selective activity of the Lunacy Commissioners) has seen to it that the correspondence is sufficient for practical needs. But we cannot relay on the correspondence, and in physics we do not accept any detail of the picture unless it is confirmed by more exact methods of inference.

The external world of physics is thus a universe populated with *inferences*. The inferences differ in degree and not in kind. Familiar objects which I handle are just as much inferential as a remote star which I infer from a faint image on a photographic plate, or an 'undiscovered' planet inferred from irregularities in the motion of Uranus. It is sometimes asserted that electrons are essentially more

---

[1]With the view of learning what might be said from the philosophical side against the abandonment of determinism, I took part in a symposium of the Aristotelian Society and Mind Association in July 1931. Indeterminists were strongly represented, but unfortunately there were no determinists in the symposium, and apparently none in the audience which discussed it. I can scarcely suppose that determinist philosophers are extinct, but it may be left to their colleagues to deal with them.

hypothetical than stars. There is no ground for such a distinction. By an instrument called a Geiger counter, electrons may be counted one by one as an observer counts one by one the stars in the sky. In each case the actual counting depends on a remote indication of the physical object. Erroneous properties may be attributed to the electron by fallacious or insufficiently grounded inference, so that we may have a totally wrong impression of what it is we are counting; but the same is equally true of the stars.

In the universe of inferences, past, present, and future appear simultaneously, and it requires scientific analysis to sort them out. By a certain rule of inference, namely, the law of gravitation, we infer the present or past existence of a dark companion to a star; by an application of the same rule of inference we infer the existence on Aug. 11, 1999, of a configuration of the sun, earth, and moon, which corresponds to a total eclipse of the sun. The shadow of the moon on Cornwall in 1999 is already in the universe of inference. It will not change its status when the year 199 arrives and the eclipse is observed; we shall merely substitute one method of inferring the shadow for another. The shadow will always be an inference. I am speaking of the object or condition in the external world which is called a shadow; our perception of darkness is not the physical shadow, but is one of the possible clues from which its existence can be inferred.

Of particular importance to the problem of determinism are our inferences about the past. Strictly speaking, our direct inferences from sight, sound, touch, all relate to a time slightly antecedent; but often the lag is more considerable. Suppose that we wish to discover the constitution of a certain salt. We put it in a test tube, apply certain reagents, and ultimately reach the conclusion that it *was* silver nitrate. It is no longer silver nitrate after our treatment of it. This is an example of retrospective inference; the property which we infer is not that 'being X' but of 'having been X'.

We noted at the outset that in considering determinism the alleged causes must be challenged to produce their birth certificates so that we may know whether they really were pre-existing. Retrospective inference is particularly dangerous in this connection because it involves antedating a certificate. The experiment above mentioned certifies the chemical constitution of a substance, but the date we write on the certificate is earlier than the date of the experiment. The antedating is often quite legitimate; but that makes the practice all the more dangerous, it lulls us into a feeling of security.

RETROSPECTIVE CHARACTERS

To show how retrospective inference might be abused, suppose that there were no way of learning the chemical constitution of a substance without destroying it. By hypothesis a chemist would never know until after his experiment what substance he had been handling, so that the result of every experiment he performed would be entirely unforseen. Must he than admit that the laws of chemistry are chaotic? A man of resource would override such a trifling obstacle. If he were discreet enough never to say beforehand what his experiment was going to demonstrate, he might give edifying lectures on the uniformity of Nature. He puts a lighted match in a cylinder of gas, and the gas burns. "There you see that hydrogen is inflammable". Or the match goes out. "That proves that nitrogen does not support combustion". Or it burns more brightly. "Evidently oxygen feeds combustion." "How do you know it was oxygen?" "By retrospective inference from the fact that the match burned more brightly." And so the experimenter passes from cylinder to cylinder; the match sometimes behaves one way and sometimes another, thereby beautifully demonstrating the uniformity of Nature and the determinism of chemical law! It would be unkind to ask how the match must behave in order to indicate indeterminism.

If by retrospective inference we infer characters at an earlier date, and then say that those characters invariably produce at a future date the manifestation from which we inferred them, we are working in a circle. The connexion is not causation but definition, and we are not prophets but tautologist. We must not mix up the genuine achievements of scientific prediction with this kind of charlatanry, or the observed uniformities of Nature with those so easily invented by our imaginary lecturer. It is easily seen that to avoid vicious circles we must abolish purely retrospective characteristics – those which are never found as existing but always as having existed. If they do not manifest themselves until the moment that they cease to exist, they can never be used for prediction except by those who prophesy after the event.

Chemical constitution is not a retrospective character, though it is often inferred retrospectively. The fact that silver nitrate can be bought and sold shows that there is a property of *being* silver nitrate as well as of *having been* silver nitrate. Apart from special methods of determining the constitution or properties of a substance without destroying it, there is one general method widely applicable. We divide the specimen into two parts, analyze one part (destroying it if necessary), and show that its constitution *has been* X; then it is usually a fair inference that the constitution of the other part *is* X. It is sometimes argued that in this way a character inferable retrospectively must always be also inferable contemporaneously; if that were true, it would remove all danger of using retrospective inference to invent fictitious characters as causes of the events observed. Actually the danger arises just at the point where the method of sampling breaks down, namely, when we are concerned with characteristics supposed to distinguish one individual atom from another atom of the same substance; for the individual atom cannot be divided into two samples, one to analyze and one to preserve. Let us take an example:

It is known that potassium consists of two kinds of atoms, one kind being radioactive and the other inert. Let us call the two kinds K_α and K_β. If we observe that a particular atom bursts in the radioactive manner, we shall infer that it was a K_α atom. Can we say that the explosion was predetermined by the fact that it was a K_α and not a K_β atom? On the information stated there is no justification at all; K_α is merely an antedated label which we attach to the atom when we see that it has burst. We can always do that, however undetermined the event may be which occasions the label. Actually, however, there is more information which shows that the burst is not undetermined. Potassium is found to consist of two isotopes of atomic wights 39 and 41; and it is believed that 41 is the radioactive kind, 39 being inert. It is possible to separate the two isotopes and to pick out atoms known to be K^{41}. Thus, K^{41} is a contemporaneous character, and can legitimately predetermine the subsequent radioactive outburst; it replaces K_α which was a retrospective character.

So much for the fact of outburst; now consider the time of outburst. Nothing is known as to the time when a particular K^{41} atom will burst except that it will probably be within the next thousand million years. If, however, we observe that it bursts at a time t, we can ascribe to the atom the retrospective character K^t, meaning that it had (all along) the property that it was going to burst at time t. Now, according to modern physics, the character K^t is not manifested in any way – is not even represented in our mathematical description of the atom – until the time t when the burst occurs and the character K^t, having finished its job, disappears. In these circumstances K^t is not a predetermining cause. Our retrospective labels and characters add nothing to the plain observational fact that the burst occurred without warning at the moment t; they are merely devices for ringing a change on the tenses.

The time of break-up of a radioactive atom is an example of extreme indeterminism; but it must be understood that, according to current theory, all future events are indeterminate in greater or lesser degree, and differ only in the margin of uncertainty. When the uncertainty is below our limits of measurement, the event is looked upon as practically determinate; determinacy in this sense is relative to the refinement of our measurements. A being accustomed to time on the cosmic scale, who was not particular to a few hundred million years or so, might regard the time of break-up of the radioactive atom as practically determinate. There is one unified system of secondary law throughout physics and a continues gradation from phenomena predictable with overwhelming probability to phenomena which are altogether indeterminate.

CRITICISM OF INDETERMINISM

In saying that there is no contemporaneous characteristic of the radioactive atom determining the date at which it is going to break up, we mean that in the picture of the atom as drawn in present-day physics no such characteristic appears; the atom which will break up in 1960 and the atom which will break up in the year 150,000 are drawn precisely alike. But, it will be said, surely that only means that the characteristic is one which physics has not yet discovered; in due time it will be found and inserted in the picture either of the atom or its environment. If such indeterminacy were exceptional, that would be the natural conclusion, and we should have no objection to accepting such an explanation as a likely way out of a difficulty. But the radioactive atom was not brought forward as a difficulty; it was brought forward as a favorable illustration of that which applies in greater or lesser degree to all kinds of phenomena. There is a difference between explaining a way an exception and explaining away a rule.

The persistent critic continues: "You are evading the point. I contend that there are characteristics unknown to you which completely predetermine not only the time of break-up of the radioactive atom but also all physical phenomena. How do you know there are not? You are not omniscient."

The curious thing is that the determinist who takes this line is under the illusion that he is adopting a more modest attitude in regard to our scientific knowledge than the indeterminist. The indeterminist is accused of claiming omniscience. I will not make quite the same countercharge against the determinist; but surely it is only a man who thinks himself *nearly* omniscient who would have the audacity to start enumerating all the things which which (it occurs to him) might exist without his knowing it. I am so far from omniscient that my list would contain innumerable entries. If it is any satisfaction to the critic, my list does include deterministic characters – along with Martian irrigation works, ectoplasm, etc. – as things which might exist unknown to me.

It must be realized that determinism is a positive assertion about the behaviour of the universe. It is not sufficient for the determinist to claim that there is no fatal objection to his assertion; he must produce some reason for making it. I do not say he must prove it, for in science we are ready to believe things on evidence falling short of strict proof. If no reason for asserting it can be given it collapses as an idle speculation. It is astonishing that even scientific writers on determinism advocate it without thinking it necessary to say anything in its favour, merely pointing out that the new physical theories do not actually disprove determinism. If that really represents the status of determinism, no reputable scientific journal would waste space over it. Conjectures put forward on slender evidence are the curse of science; a conjecture for which there is no evidence at all is an outrage. So far as the physical universe is concerned, determinism appears to explain nothing; for in the modern books which go farthest into the theory of the phenomena no use is made of it.

Indeterminism is not a positive assertion. I am an indeterminist in the same way that I am an anti-moon-is-made-of-green-cheese-ist. That does not mean that I especially identify myself with the doctrine that the moon is *not* made of green cheese. Whether or not the green cheese lunar theory can be reconciled with modern astronomy is scarcely worth inquiring; the main point is that green-cheesism, like determinism, is a conjecture that we have no reason for entertaining. Undisprovable hypotheses of that kind can be invented *ad lib*.

Principle of Uncertainty

The mathematical treatment of an indeterminate universe does not differ much in from from the older treatment designed for a determinate universe. The equations of wave mechanics used in the new theory are not different in principle from those of hydrodynamics. The fact is that, since an algebraic symbol can be used to represent either a known or an unknown quantity, we can symbolize a definitely predetermined future or an unknown future in the same way. The difference is what whereas in the older formulae every symbol was theoretically determinable by observation, in the present theory there occur symbols the values of which are not assignable by observation.

Hence, if we use the equations to predict, say, the future velocity of an electron, the result will be an expression containing, besides known symbols, a number of undeterminable symbols. The latter make the prediction indeterminate. (I am not here trying to prove or explain the indeterminacy of the future; I am only stating how we adapt our mathematical technique to deal with an indeterminate future.) The indeterminate symbols can often (or perhaps always) be expressed as unknown phase-angles. When a large number of phase-angles are involved, we may assume in averaging that they are uniformly distributed from 0^0 to 360^0, and so obtain predictions which could only fail if there has been an unlikely coincidence of phase-angles. That is the secret of all our successful prophecies; the unknowns are not eliminated by determinate equations but by averaging.

There is a very remarkable relation between the determined and the undetermined symbols, which is known as Heisenberg's Principles of Uncertainty. The symbols are paired together, every determined symbol having an undetermined symbol as partner. I think that this regularity makes it clear that the occurrence of undetermined symbols in the mathematical theory is not a blemish; it gives a special kind of symmetry to the whole picture. The theoretical limitation on our power of predicting the future is seen to be systematic, and it cannot be confused with other casual limitations due to our lack of skill.

Let us consider an isolated system. It is part of a universe of inference, and all that can be embodied in it must be capable of being inferred from the influence which it broadcasts over its surroundings. Whenever we state the properties of a body in terms of physical quantities, we are imparting knowledge as to the response of various external indicators to its presence and nothing more. A knowledge of the response of all kinds of objects would determine completely its relation to its environment, leaving only its unget-at-able inner nature, which is outside the scope of physics. Thus, if the system is really isolated so that it has no interaction with its surroundings, it has no properties belonging to physics, but only an inner nature which is beyond physics. So we must modify the conditions a little. Let it for a moment have some interaction with the world exterior to it; the interaction starts a train of influences which may reach an observer; he can from this one signal draw an inference about the system, that is, fix the value of one of the symbols describing the system or fix one equation for determining their values. To determine more symbols there must be further interactions, one for each new value fixed. It might seemed that in time we could fix all the symbols in this way, so that there would be no undetermined symbols in the description of the system. But it must be remembered that the interaction which disturbs the external world by a signal also reacts on the system.

There is thus a double consequence; the interaction starts a signal through the external world informing us that the value of a certain symbol p in the system is p_1, and at the same time it alters to an indeterminable extent the value of another symbol q in the system. If we have learned from former signals that the value of q was q_1, our knowledge will cease to apply, and we must start again to find the new value of q. Presently there may be another interaction which tells us that q is now q_2; but the same interaction knocks out the value p_1 and we no longer know p. It is of the utmost importance for prediction that a paired symbol and not the inferred symbol is upset by the interaction. If the signal taught us that at the moment of interaction p was p_1, but that p had been upset by the interaction and the value no longer held good, we should never have anything but retrospective knowledge – like the chemistry lecturer to whom

I referred above. Actually we can have contemporaneous knowledge of the values of half the symbols, but never more than half. We are like the comedian picking up parcels who, each time he picks up one, drops another.

There are various possible transformations of the symbols and the condition can be expressed in another way. Instead of two paired symbols, one wholly known and the other wholly unknown, we can take two symbols each of which is known with some uncertainty; then the rule is that the product of the two uncertainties is fixed. Any interaction which reduces the uncertainty of determination of one increases the uncertainty of the other. For example, the position and velocity of an electron are paired in this way. We can fix the position with a probable error of about of 0.001 mm. and the velocity with a probable error of about 1 km. per sec.; or we can fix the position to 0.0001 mm. and the velocity to 10 km. per sec.; and so on. We divide the uncertainty how we like, but we can not get rid of it. If current theory is right, this is not a question of lack of skill or a perverse delight of Nature in tantalizing us; for the uncertainty is actually embodied in the theoretical picture of the electron; so that if we describe something as having exact position and velocity we cannot be describing an electron.

If we divide the uncertainty in position and velocity at time t_1 in the most favorable way, we find that the predicted position of the electron one second later (at time t_2) is uncertain to about five centimeters. That represents the extent to which the future position is not predetermined by anything existing one second earlier. If the position at time t_2 always remained uncertain to this extent, there would be no failure of determinism, for the thing we had failed to predict (exact position at time t_2) would be meaningless. But *when the second has elapsed* we can measure the position of the electron to 0.001 mm. or even more closely, as already stated. This accurate position is not predetermined; we have to wait until the time arrives and then measure it. It may be recalled that the new knowledge is acquired at a price. Along with our rough knowledge of position (to 5 cm.) we had a fair knowledge of the velocity; but when we acquire more accurate knowledge of the position, the velocity goes back into extreme uncertainty.

We might spend a long while admiring the detailed working of this cunning arrangement by which we are prevented from finding out more than we ought to know. But I do not think we should look on these as Nature's devices to prevent us from seeing too far into the future. They are the devices of the mathematician who has to protect himself from making impossible predictions. It commonly happens that when we ask silly questions, mathematical theory does not directly refuse to answer but gives a non-committal answer like $\frac{0}{0}$, out of which we cannot wring any definite meaning. Similarly, when we ask where the electron will be tomorrow, mathematical theory does not give the straightforward answer, "It is impossible to say, because it is not yet decided", because that is beyond the resources of an algebraic vocabulary. It gives us an ordinary formula of x's and y's, but makes sure that we cannot possibly find out what the formula means – until tomorrow.

MENTAL INDETERMINISM

I have, perhaps fortunately, left myself no time to discuss the effect of indeterminacy in the physical universe on our general outlook. I will content myself with stating in summary form the points which seem to arise.

(1) If the whole physical universe is deterministic, mental decisions (or at least *effective* mental decisions) must also be predetermined. For if it is predetermined in the physical world (to which your body belongs) that there will be a pipe between your lips on Jan. 1, the result of your mental struggle on Dec. 31 as to whether you will give up smoking in the New Year is evidently predetermined. The new physics thus opens the door to indeterminacy of mental phenomena, whereas the old deterministic physics bolted and barred it completely.

(2) The door is opened slightly, but apparently the opening is not wide enough; for according to analogy with inorganic physical systems, we should expect the indeterminacy of human movements to be quantitatively insignificant. In some way we must transfer to human movements the wide indeterminacy characteristic of atoms, instead of the almost negligible indeterminacy manifested by inorganic systems of comparable scale. I think this difficulty is not insuperable, but it must not be underrated.

(3) Although we may be uncertain as to the intermediate steps, we can scarcely doubt what is the final answer. If the atom has indeterminacy, surely the human mind will have an equal indeterminacy; for we can scarcely accept a theory which makes out the mind to be more mechanistic than the atom.

(4) Is the human will really more free if its decisions are swayed by new factors born from moment to moment than if they are the outcome solely of heredity, training, and other predetermining causes? On such questions as these we have nothing new to say. Argument will no doubt continue 'about it and about'. But it seems to me that there is a far more important aspect of indeterminacy. It makes it possible that the mind is not utterly deceived as to the mode in which its decisions are reached. On the deterministic theory of the physical world, my hand in writing this address is guided in a predetermined course according to the equations of mathematical physics; my mind is unessential – a busybody who invents an irrelevant story

about a scientific argument as an explanation of what my hand is doing – an explanation which can only be described as a downright lie. If it is true that the mind is so utterly deceived in the story it weaves round our human actions, I do not see where are we to obtain our confidence in the story it tells of the physical universe.

Physics is becoming difficult to understand. First relativity theory, then quantum theory, then wave mechanics have transformed the universe, making it seem ever more fantastic to our minds. Perhaps the end is not yet. But there is another side to this transformation. Naïve realism, materialism, the mechanistic hypotheses were simple; but I think that it was only by closing our eyes to the essential nature of experience, relating as it does to the reactions of a conscious being, that they could be made to seem credible. These revolutions of scientific thought are clearing up the deeper contradictions between life and theoretical knowledge, and the latest phase with its release from determinism marks a great step onwards. I will even venture to say that in a present theory of the physical universe we have at last reached something which a reasonable man might almost believe.

80

Appendix: The Philosophical Aspect of the Theory of Relativity: A Symposium

A. S. Eddington, W. D. Ross, C. D. Broad and F. A. Lindemann, The Philosophical Aspect of the Theory of Relativity: A Symposium, *Mind* **29** (116), 415-445 (1920). Contributed to the International Congress of Philosophy, 1920.

I. By A. S. Eddington

It is natural for a scientific man to approach Einstein's theory of Relativity with some suspicion, looking on it as an incongruous mixture of speculative philosophy with legitimate physics. There is no doubt that it was largely suggested, by philosophical considerations, and it leads to results hitherto regarded as lying in the domain of philosophy and metaphysics. But the theory is not, in its nature or in its standards, essentially different from other physical theories; it deals with experimental results and theoretical deductions which naturally arise from them. The only point in which it shocks our conservatism is that it regards the investigation of the properties of physical time and space as being a legitimate subject of experimental and theoretical research, like the investigation of the properties of matter. Time and space are things which a physicist is continually using and measuring; and it is difficult to see why he should not be allowed to investigate their properties without being condemned as a metaphysicist. I think the opposition arises from the impression that in their physical aspects the properties of time and space are so simple and so inevitable that we have long known all that there is to be learnt by physical methods; and therefore if an investigator spends any time over these he must necessarily be trespassing beyond legitimate physics. On the other hand, we know that much remains to be found out as to the physical constitution of matter; and so the man who occupies himself with it is not presumed to be speculating metaphysically as to the meaning of substance. But the relativity theory makes it clear that the experimental study of the physical aspects of space and time has not been exhausted; it applies the recognised scientific method to this study; and there is no breach of continuity with ordinary physics. It unfolds a physical theory of space and time and matter, which, we can scarcely doubt, marks a great advance. It would be rash to suppose that it reaches finality; but it bears all the indications of being one of the more permanent stages in the advance towards Truth.

I would emphasise then that the theory of relativity of time and space is essentially a physical theory, like the atomic theory of matter or the electromagnetic theory of light; and it does not overstep the natural domain of physics. But, speaking to an audience of philosophers, I shall not hesitate to trespass beyond the borderline on my own account. I shall be a stranger in a strange country; and the lurking pits might well intimidate me, if I did not rely on your friendly hands to pick me out.

We can perhaps obtain some insight into the meaning of Relativity by analysing the idea of "green." Green light is primarily a sensation experienced by a normal individual, which is obviously subjective. In current physics it is supposed that there is in the external world an exact objective counterpart to green light, viz., electromagnetic oscillations of a particular quantitative character; and, so far as physics is concerned, the name "green light" is transferred to this objective counterpart. Further this quantitative character can be consistently estimated by physical appliances other than the eye, so that even in its subjective aspect it is no longer necessary to insist on the psychological significance of green. We ought now to be able to dispense with the idea of any recipient of the light, so that there are electromagnetic waves in Nature which can be described as absolutely green. But that is too hasty a conclusion. If we take an observer travelling rapidly to meet these waves, they will appear to him not green but blue; if this is an illusion, it is shared by his spectroscope, his photo-electric cell, the chlorophyl of the plants, by everything travelling with him. For a whole moving world the light is blue; for a differently moving planet it will be orange; what meaning then can we attach to its absolute greenness? Why have we singled out green as the true colour, when to the different conceivable worlds it takes all hues of the rainbow? We are foiced to admit that we called it green merely because it was green for some particular observer whom we had in mind at the start. Now here modern experimental investigation comes in; we have entirely failed to discover anything pre-eminent about this particular observer, or any other observer, entitling his views to more weight than those of observers with different motions. If we lost him, there is no criterion whatever by which we could reconstruct him. It is the old philosophical point (perhaps unexpectedly applicable) that absolute motion is meaningless and undetectable, and therefore observers merely differing in their motions present no criterion for singling out a leader.

We cannot call the light absolutely green, when it is only green for a particular observer arbitrarily selected. This drives us back practically to the starting point; green is not an objective quality of the light. Even when we have abstracted the psychological significance of colour, it still remains a relation on which the objective reality and some specified recipient are both involved. It is commonly said that a sodium atom always radiates yellow light; but the light is only yellow relative to the atom itself, or to an observer having the same motion. Intrinsically the light has no particular colour, and observers can be imagined for whom it is violet or red. The relativity theory does not arbitrarily divide this colour into objective yellowness plus a correction for the motion of the recipient; it simply accepts the plain fact that the colour-name applies to a relation of the reality to a recipient.

At first sight this seems to throw over the common view that colour is determined by the length of the electromagnetic waves. Is not the true and absolute colour-quality that which corresponds to the length of the waves; whereas the colour actually perceived may be modified by the observer's motion according to well-known principles? This brings us to the most revolutionary idea in the relativity theory. Length itself is not an absolute character intrinsic in the external world; like colour, it is a relation between the thing in Nature and the observer, being modified by his motion. This has escaped common notice, because all observers who can compare notes share practically the same motion – that of the earth. It is only recent delicate experiments that have revealed it. If length cannot be relied on as absolute, what shall we say of the other quantities of physics? The answer comes that all the more familiar terms of physics – duration of time, mass, force, energy, etc. – denote not objective characters, but relations to some observer or his idealised equivalent; and, in particular, these relations are modified by his motion.

We thus see that the knowledge contained in current physics is only a knowledge of the relations of Nature to particularly circumstanced observers. It is not on that account to be condemned; we shall continue to study and extend this relative knowledge. But it is important in many cases in physics, and still more in philosophy, to appreciate its relativity. We must make a special study of the way in which the relation changes for differently circumstanced observers, and abandon the crude methods which arose under the mistaken impression that under the familiar names we were dealing with things objective and independent of us. When this is done many of the perplexities of modern science are cleared up, and a great simplification results.

Since physics has not hitherto dealt with the absolute world, we may ask whether it is competent to do so. It is. The problem is not so very difficult to solve; it was not solved before because until recently we were unaware that there remained such a problem to solve. To put the claim rather more modestly and more accurately, we can arrive at a description of the physical phenomena which is independent of the motion of the observer (that being apparently the confusing factor in our present relative knowledge). In a sense the expression of this knowledge is still relative, because our imaginations can only work with material which is in some degree familiar; but the recipient, whom we set up to relate external Nature to, is now only a dummy whom we can change freely without altering anything in the description. It is not like the older relative knowledge in which *green* has to become *red* when we change the observer.

The absolute world of physics thus reached is four-dimensional, events outside us being arranged in an indissoluble four-fold order which may be regarded as a combination of space and time. Space and time are relations to an individual, and as relations are quite separate. But there is not one objective reality at the far end of the space-relation, and another reality at the far end of the time-relation; both relations spring from one common source. Perhaps I may venture to indicate how the common distinction of space and time arises. The observer himself is part of the world, and from a four-dimensional point of view we must regard him as having the form of a *worm*. He distinguishes the order of events in the direction of his length as time, and his other three dimensions he regards as space. He applies this to his own elongated form, and considers that he himself has considerable duration in time, but more modest extension in space. We easily see that worms whose lengths lie in different directions (or, as we should ordinarily say, individuals who are moving with different velocities) make a different dissection into time and space. But this is not all; the objective four-dimensional continuum is indissoluble; but if we take in it two arbitrary events A and B, the relation between them (out of which their physical aspects arise) is one or other of two qualitatively distinct kinds. On developing the theory, it is found that if the relation of A to B is of the first kind it is possible for a particle of matter to extend from A to B, but not if it is of the second kind. That is a property inherent in the constitution of matter. In physics we deal only with observers who possess material bodies, however abstract they may be in other respects; and consequently the length of one of our worms cannot lie along AB unless the relation between the two points is of the first kind. (In ordinary language the observer must not travel faster than light.) It follows that although the worms can lie in all kinds of directions within wide limits, yet in every case the relations of events along the length of a worm, which he takes to be the time-order, are qualitatively and objectively of a different kind from the relations in transverse directions which he adopts as space. That is why time and space appear and are so different. The observer's velocity (or four-dimensional extension) determines his separation of time and space; but behind that there is a rudimentary objective differentiation of orderly relation, which limits the observer's velocity and is by that means carried through into the resulting separation.

We believe that this theory (or rather the analysis which is equivalent to it) greatly elucidates the meaning of our measurements of space and time, and has far reaching consequences in physics. I doubt whether its importance in philosophy is so immediate as is often supposed, because it leaves us still with an objective distinction between time-like and space-like order. The mathematician differentiates these by the aid of his symbol $\sqrt{-1}$; but that, of course, does not throw light on their intrinsic unlikeness.

Minkowski summed up the earlier relativity theory in the celebrated phrase, "Time and Space in themselves sink to mere shadows." Moritz Schlick, in his admirable book,[1] has said that this must now be extended to Time and Space and Things sink to shadows. "The combination or oneness of space, time and things is alone reality; each by itself is an abstraction." With *things* I take it that he includes not only matter but all that is commonly supposed to be *in* space and time, for example, fields of force. It is so easy to give glib acceptance to this doctrine, so difficult to rise to it in our outlook on physics. The non-Euclidean heterogeneous space of Einstein is a natural consequence of this view; for "things" are everywhere heterogeneous, and it is unlikely that the same oneness can manifest itself as homogeneity in its space-aspects and heterogeneity in its thing-aspects.

I have tried to show elsewhere[2] the exact method by which, starting from a relation undefinable in its absolute character, we arrive from a single source at the physical quantities which describe space and time on the one hand and the quantities which describe things on the other hand. If we describe the character (or geometry) of space and time throughout the world, we at the same time necessarily describe all the things in the world. The conspicuous instance of this is in Einstein's theory of gravitation, where in describing the geometry of space and time throughout the solar system, he finds himself describing at the same time the sun's gravitational field. The same applies also to other things such as matter. The difference between space occupied by matter and space which is empty is simply a difference in its geometry. There seems to be no reason to postulate that there is an entity of foreign nature present which causes the difference of geometry; and if we did postulate such an entity it would scarcely be proper to regard it as physical matter, because it is not the foreign entity but the difference of geometry which is the subject

---

[1] *Space and Time in Contemporary Physics.*
[2] MIND, Vol. XXIX., No. 114.

of physical experiment.

In contemplating the starry heavens, the eye can trace patterns of various kinds – triangles, chains of stars, and more fantastic figures. In a sense these patterns exist in the sky; but their recognition is subjective. So out of the primitive events which make up the external world, an infinite variety of "patterns" can be formed. There is one type of pattern which for some reason the mind loves to trace wherever it can; where it can trace it, the mind says, "Here is substance;" where it cannot, it says "How uninteresting! There is nothing in my line here." The mind is dealing with a real objective substratum; but the distinction of substance and emptiness is the mind's own contribution, depending on the kind of pattern it is interested in recognising. It seems probable that the reason for selecting the particular type of pattern is that this pattern has (from its own geometrical character, and independently of the material in which it is traced) a property known as *Conservation*. Reverting from the four-dimensional world to ordinary space and time, this property appears as *permanence*. That the mind would necessarily choose for the substance of its world something which is permanent seems natural and inevitable. The interesting point is that there is no obligation on Nature to provide explicitly anything permanent; the permanence is introduced by the geometrical quality of the configuration, which the mind looks out for in whatever Nature provides.

Now it appears that a great number of the well-known laws of physics, mechanics and geometry are implicitly contained in this identification of substance. That is to say, these laws do not govern the course of events in the objective world, but are automatically imposed by the mind in selecting what it considers to be substance. They are identities contained in the definition of the geometrical character of the pattern which the mind hunts out. If all the discoveries of physics related to laws of this kind, we should be forced to admit that physics has nothing to contribute to the great question of how the world outside us is governed. I am not as yet prepared to admit that. I think that we do, more especially in modern physics, encounter the genuine laws governing the external world, and are attempting – perhaps rather unsuccessfully – to grapple with them. But the great exact laws of gravitation, mechanics and electromagnetism, by which physics has won its high reputation as an exact science, all appear to belong to the other category; and, when these are set aside as irrelevant, our claim to have grasped the type of law, or even the meaning of law, prevailing in the world outside us is reduced to very modest proportions.

An aged college-bursar once dwelt secluded in his rooms devoting himself entirely to accounts. He had cut himself off entirely from the life around him, and he realised the intellectual and other activities of the college only as they reflected themselves in the bills. The accounts were his world; and the different items took on an individuality in his mind. He vaguely pictured an objective reality at the back of it all – some sort of parallel to the real college – though he could only imagine it in terms of £. s. d., which constituted its relation to him. His method of account-keeping had become inevitable habit, handed on to him from a long succession of hermit-like bursars; and he had no idea that he was in any way concerned in the method; it seemed impossible that the accounts could be put in any other way. But he was of a scientific turn, and he wanted to know more about the college – the world of his accounts. One day, in looking over the books, he discovered a remarkable thing. For every item which appeared on the credit side of the account, an equal item appeared somewhere else on the debit side. "Ha!" said the bursar, "I have discovered one of the great laws governing the college. It is a perfect exact law of nature with no exceptions. Credit must be called plus and debit minus; so we have the law of conservation of £. s. d. This is the mode of investigation which alone can give me sure knowledge of the world, and I see no limits to the field it will ultimately cover. I have only to go on in this way, and I shall begin to understand why it is that prices are always going up."

Perhaps it is conservatism, but I am not prepared to press this analogy quite to its apparent conclusion. I do think that we have, like the bursar, tended to confuse the laws of economics with the laws of accounts – the laws under which the objective world is developing itself, and the laws inherent in the overlapping of the different aspects under which we relate it to ourselves. I think that the results in which physics has been so conspicuously successful are mainly of the latter character. But I think that the bursar's method of investigation was a sound one; I would not have him give up his books, and turn in despair to the faint confused sounds of an outside activity which from time to time penetrate the walls of his cell. *Ne sutor ultra crepidam.* The laws of economics are not going to be reached so easily as he supposed; they are not even on the same plane as his first sensational discovery belonged to; but by diligent study of his world of accounts he may yet be able to puzzle out something of the activity behind.

And so, when the seed reproduces the character of its parent, when the tree clothes itself in leaves, when philosophers are drawn together in congress, it may be misleading to compare the motive-laws with the familiar type illustrated by the law of gravitation. The line of demarcation is not between vital and inert phenomena. The point is that the idea of law even in the world of inert matter may, in some way as yet undefined, transcend the instances which are as yet known; that these instances are, indeed, not fair parallels for comparison. The old type of law must, of course, always be obeyed – the college may totter, but the bursar's accounts still balance. If this is indeed so, it will not be easy for the physicist, who, however, has already a strong suspicion that in the quantum phenomena, which he is now encountering everywhere, he is up against laws of a different type from those which have hitherto succumbed to his inquiries. But in the wider outlook on life this emancipation, if it prove true, is likely to be hailed with relief.

II. By W. D. Ross

I do not propose in my contribution to the Symposium to discuss Prof. Eddington's paper, interesting as it is. His paper gives us not the line of argument which leads up to the theory of relativity, but rather the further speculations of one who has already been convinced by that line of argument. My difficulties begin further back, with the argument itself, and it is to some aspects of it that I will address myself. I should like, however, to comment on two remarks of his. 'There is no doubt,' he says, that Einstein's theory 'was largely suggested by philosophical considerations,' and a little later, 'It is the old philosophical point ... that absolute motion is meaningless and undetectable.' It seems to be supposed by many of the scientists who have discussed the subject that philosophy condemns absolute motion, apart from any of the experimental grounds on which they themselves reject it; and they feel themselves fortified by this support from an independent source. Many philosophers have no doubt rejected absolute motion, but many others believe in it. For my own part, I think that Mr. Russell's chapter on the subject[3] is a complete refutation of at any rate the main philosophical arguments that have been urged against absolute motion.

I would make one other preliminary remark, with reference to Prof. Eddington's first page. The division of opinion about Einstein's theory is not in any sense one in which science and philosophy are ranged on opposite sides. Both scientists and philosophers are divided on the question; and the truth, on whichever side it lies, is to be reached by close thinking on certain questions, in one sense very simple, in another extremely difficult, competence to discuss which is not the monopoly of either scientists or philosophers, but whose solution is not so easy that either class of thinkers can afford to reject the aid of the other.

One of the difficulties about relativity is that its supporters seem in the very act of arguing for it to be implying its opposite. I will confine myself to the 'special theory;' until one can be satisfied about the truth of this, it would be useless to discuss the general theory which is an extension and in some degree a correction of it. Incidentally, one's faith in the argument, should surely be somewhat shaken by the fact that the constant relative velocity of light, which is asserted in the special theory, is denied in the general. Were it a question

---

[3]In *Principles of Mathematics*

of getting nearer to the truth by further experiment, there would be nothing surprising in this; but it is not satisfactory that 'the keystone of the old theory'[4] should later be so cheerfully dispensed with.

It seems to be generally agreed among relativists that the theory is forced on us in the first instance by the result of Michelson-Morley's experiment. We naturally assume light to have a constant absolute velocity in all directions; we therefore expect its velocity relative to the earth to be affected by the motion of the earth; but we find that apparently it is not. Hence we seem to be driven to accept one or other of two surprising theories, that of Lorentz or that of Einstein. Now why should we not adopt the hypothesis that the earth is at rest relatively to the ether? If it is, we should expect rays of light moving in different directions above the earth's surface to move with constant velocity relative to a starting-point on the earth, and there would be nothing surprising in the result of the experiment. But, I shall be told, this is to go back to the Ptolemaic view, which has long since been exploded. This, however, is not *my* solution; I am simply asking why it should not be the solution for a disbeliever in absolute motion. According to him, it is just as true that the station moves past the train as that the train moves past the station. It is then, as true that the rest of the universe moves, relatively to the earth as that the earth moves relatively to the rest of the universe. The Copernican view is no truer than the geocentric; in fact they are the same view. 'But neither Ptolemy nor Copernicus was really right,' relativists will say, 'neither the earth nor the remainder of the universe is at rest; both are in relative motion, which is the only motion there is, and it is the existence of this motion that makes the Michelson-Morley result surprising and Einstein's explanation of it necessary.' But it is *not* the motion of the earth relative to the stars that makes the result surprising; it is the presumed motion of the earth relative to the ether. Now, that no such motion can be detected is a fundamental principle of their theory. Why, then, assume, as they do in their whole consideration of the experiment, that such motion exists? On their principles, the relative motion of the earth and the stars only requires that *one of the two* should be in motion relatively to the ether. The assumption that it is the earth that is so shows that relativists are Copernicans, and therefore at bottom not relativists.

But, I may now be told, relativists do not believe in an ether at all. They speak with a divided voice on the subject, but their general opinion seems to be against this unfortunate entity, whose alleged attributes have always somewhat scandalised philosophers. There is, then, only motion of ordinary bodies relatively to one another. But then there is nothing whatever in the Michelson-Morley result

---

[4]I.e., of the 'special theory.' Prof. Broad in *Hibbert Journal*, April, 1920, p. 426.

to surprise and to call for Einstein's theory. There is no reason why the motion of the earth relative to the heavenly bodies should affect the velocity of rays of light in a laboratory, which have nothing to do with the heavenly bodies but only with the earth. It is only the assumption that the earth is moving (a) absolutely, or (b) at least with regard to the ether, that makes the result surprising and calls for either the Lorentz or the Einstein explanation. Disbelievers in absolute motion and in the ether have no need of Einstein's theory, and believers in absolute motion cannot accept it because it denies absolute motion.

Take, again, another assumption which is made by relativists in discussing the Michelson-Morley result. Prof. Broad[5] states three assumptions, and says that 'the rejection of any of them will merely bring us into conflict with some other set of well-attested experimental facts.' It is on the basis of the acceptance of these assumptions that all solutions other than those of Lorentz and Einstein are ruled out. One of these assumptions is 'that the velocity of light in stagnant ether is the same in all directions.' This assumption is described as 'the only reasonable one to make on the subject,' and it is rightly pointed out that its rejection would land us in greater difficulties than its acceptance involves. This does not mean that the *relative* velocity of light is constant. For this is the conclusion which is supposed to be established by the experiment, and therefore must not be presupposed in considering what is to be deduced from the result of the experiment. As far as I can see (though I may very well be mistaken) it can only mean (1) that light moves in equal times over equal distances in space, irrespective of direction, or (2) that it moves with equal velocity relatively to bodies at rest (or in like motion) relatively to the source of light, but in different directions from it. On the first interpretation, absolute motion is already admitted in one of the assumptions on which the proof of relativity rests. This interpretation will of course be rejected, and we come to the second. Suppose then that one of the bodies which are at rest relatively to the source of light begins to move towards it. Then the velocity of light relatively to it will become greater than its velocity relatively to the bodies that are still at rest relatively to the source. For Einstein, though he rejects Newton's addition-formula for velocities, sets up another in its stead; when two velocities are added the result is something different from either, though not (as Newton said) the arithmetical sum of the two. Therefore the velocity of light relatively to two bodies, one moving towards the source of light, and the other at rest with respect to it, will be different. Thus the assumption on which the argument rests is inconsistent with the statement in the theory, that the velocity of light relatively to all

---

[5]Prof. Broad in *Hibbert Journal*, April, 1920, pp. 427, 428.

bodies is unaffected by their motion.[6]

Let me take a further illustration of the inconsequence which seems to beset even the acutest thinkers when under the influence of the glamour of relativity. Einstein[7] makes the assumption that two points of a railway line have been struck by lightning, and asks whether the statement that the strokes were simultaneous has any meaning. The reader is supposed to reply that the meaning is clear, but that he would find it difficult to say whether the statement' was true. Einstein is not satisfied with this answer. 'A concept does not exist for the physicist until the possibility of discovering in the concrete case whether the concept applies or not is given.' The question how you could possibly discover the applicability or non-applicability of a concept that does not exist for you either does not occur to Einstein, or is deemed unworthy of notice; and it is inferred that in order to have a conception of simultaneity at all we need such a definition of it that we can determine whether the lightning strokes were simultaneous. The definition proposed is that the strokes are simultaneous if they are perceived simultaneously by an observer placed midway and furnished with an apparatus (e.g., two mirrors placed at right angles) which allows a simultaneous optical fixation of the points A and B which were struck. The definition is obviously circular, and it becomes clear that what Einstein is looking for is not a definition but a test, and a test not of simultaneity, but of the simultaneity of two events not directly observed; for the test evidently rests on the observer's immediate judgment of the simultaneity of two events in his own consciousness. Thus it is clear that we have a conception of simultaneity before we set up the criterion which according to Einstein first gives us that conception. And, further, it is clear that we mean the same thing by 'simultaneous,' whether we are speaking of events in our own consciousness or of events without it, though for the application of the word in the latter case we need a criterion which we did not need before applying it in the former.

Einstein supposes the above criterion to be met by the following criticism: 'I cannot tell whether light propagates itself with the same velocity from A to M and from B to M unless I already have at my disposal the means of measuring time; the reasoning therefore is circular.' His reply is: 'My definition makes no assumption about light. The definition of simultaneity has only to be such that in every real case it enables us to decide empirically whether the concept to be defined is applicable. That light takes the same time to travel both these journeys is not an assumption about the physical nature of light, but a statement I am free to make in order to reach a definition

---

[6]This is what is *said*; what is *meant* can surely only be that observers' estimates of its velocity are unaffected by their motion. But to distinguish the tact from the estimates of it is to give up relativity.

[7] *Über die spezielle und die allgemeine Relativitätstheorie*, p. 14.

of simultaneity.' In other words, we have a word 'simultaneity' but we attach initially no meaning to it; we get tired of making this meaningless noise, and decide to attach some meaning to it, and a meaning such that in terms of it we shall be able to say of any two events that they are or that they are not simultaneous. The important thing is to make some decision, not to make the right decision; as the word, so far, means nothing, there is no right or wrong about it. We assume that light takes the same time to travel equal distances, but this is not to make any statement about the physical nature of light, since 'same time' is equally meaningless with 'simultaneous.' It is of course obvious that so long as we do not want to make a right decision, but merely some decision, the assumption that light takes twice as long to travel a certain distance west as to travel an equal distance east, or the assumption that all telegraph boys move with equal speed, would do just as well.

It is surely clear that Einstein's supposed reader was right, in saying that he does attach a definite meaning to 'simultaneous,' but does not always know whether two events are simultaneous; and it is clear that if he is to use light signals as a test of this he must know whether light does travel equal distances in equal times, as a matter of hard fact and not as a matter of mere arbitrary use of language. It is surprising that scientists should allow themselves to be fobbed off with the latter, which is all that on his own showing Einstein has to offer. However much he may deny it, the statement that light takes the same time to travel equal distances is a statement about the nature of light.

Take, again, the argument by which he proves the relativity of simultaneity (p. 16 ff.). He propounds the question whether events simultaneous in reference to the railway line are simultaneous in reference to a train moving along it. An observer on the line at M midway between A and B will judge the strokes of lightning simultaneous if the rays sent out from A and B at the time of the strokes reach him simultaneously. But an observer at M', the point on the train which was opposite M when the strokes (as judged from the line) occurred, will (if the train is moving towards B) be nearer to B than to A before either ray reaches him; the ray from B will therefore reach him before that from A, and he will judge the stroke at B to have happened before that at A. Thus two events which are simultaneous relatively to the line are not simultaneous relatively to the train. Hence simultaneity is relative, and any two things which are in relative motion have separate times of their own. On this argument three comments may be made.

(1) The relativity, if relativity there be, is relativity to minds, not to bodies. Leave out the judgments formed by the two observers, and the bottom drops out of the argument. This is obscured by Einstein when he describes each *body* of reference as having its separate time.

The theory is at bottom a form of the old philosophical doctrine of the relativity of our judgments to, their dependence on, the peculiarities of our own minds. The novel element in Einstein's theory is that the peculiarity of each mind on which he makes its judgments depend is its situation at a body which is in motion relatively to other bodies. The relativity is a relativity to bodies only as actual or possible situations of minds, or of the sense-organs used by minds.

(2) Not only are the 'local times' really judgments about time depending on the motion of the observer, but the discrepancy between the two observers' judgments can be removed. The observers have only to allow for their relative motion; they will then make the same judgment. To this the relativist will reply, 'that may be so in the illustration; we have there supposed the train to be in motion, and to be known to be in motion, relatively to the line; but in actual fact we are not in that position. No experiment has ever revealed whether the earth is moving through the ether, and if so, how fast. Therefore we do not know what allowance should be made for such motion; the only reasonable thing is to ignore it, to treat it as making no difference to the velocity of light relatively to us; events which are simultaneous to one observer will then necessarily be non-simultaneous to another, and simultaneity will necessarily be relative.' I think we must agree that we do not know whether or how fast we are moving, and therefore do not know what allowance to make for such movement. But surely the reasonable attitude is, not to say that we are theoretically right in making no allowance, that the conflicting judgments which will follow if we make no allowance are all of them right, and that therefore the same two events are and are not simultaneous. The reasonable thing is to say 'I do not know how much allowance should be made for my motion, but as my velocity is probably very small in comparison with that of light it will for most purposes make no difference. I will therefore ignore it. Anyhow I am just as likely to be right as if I made some arbitrary allowance.' Of conflicting judgments about simultaneity, then, certainly all but one, and perhaps all, will be wrong, but we cannot know, where the conflict depends on the unknown velocity of the earth, which, if any, is right. This seems to be the moral to be drawn, and though it is not the moral drawn by relativists, we owe it to them that it has been forced on our attention.

(3) It is surely clear that Einstein's argument to show that the two observers will make conflicting judgments rests on the assumption that the rays from A and B either start definitely at the same time or definitely at different times. In other words it is on the basis of an unacknowledged belief in absolute time that his argument here is worked out, and apart from that belief nothing whatever could be asserted about the times at which the messages will reach M and M'.

The conclusion to be drawn appears to be that the belief in absolute space, absolute time, and absolute motion is not a mere prejudice of common sense, but something that necessarily underlies all our thought, and that the argument which tries to disprove them is assuming them all the time. For the mathematical genius which has worked out the relativist view of the world we who are not mathematicians can have nothing but the profoundest admiration, but the superstructure is worthless unless the foundations are well and truly laid in general thinking about motion, distance, and simultaneity; and there are some of us who have no conviction that this has been done. Until we can be led to see our error, we are bound to think that the explanation of the Michelson-Morley and similar results is to be found in some theory not about space and time but about matter or ether, some explanation like that of Lorentz, which seems to us, though surprising enough, to contain nothing that we need have any difficulty in believing. Since its transformation equations are identical with those of Einstein, I take it that Lorentz's theory will do all the work that Einstein's special theory will do. The latter theory seems to rest on a fundamental confusion between facts and the estimates which different observers will form of them.

III. By C. D. Broad

I shall deal first with the difficulties found by Mr. Ross in arguments that have been used for the special theory of relativity. I think that these difficulties rest mainly on misunderstandings, and that they can easily be removed by a little explanation.

(i) Mr. Ross regards it as a weakness that the constancy of the velocity of light should be the keystone of the special theory and yet be discarded in the general theory. There is no real difficulty here, when we remember the different subjects with which the two theories are concerned. The special theory explicitly confined itself to systems in uniform translational motion with respect to a Newtonian frame of reference. It did not profess to tell us what would happen if a system rotated with respect to such a frame or moved with an accelerated rectilinear motion with respect to it. Now the general theory professes to deal with *all* motions, no matter to what they may be relative or what may be their kinematic characteristics. There is nothing startling in the fact that a proposition which is true and important for a restricted class of motions should not be true of all motions whatever. Mr. Ross would not, I trow, feel any difficulty if he were told that certain phonetic laws are the keystone of the sound-changes in Teutonic languages, but that they are not true without modification when we take into account all Indo-European languages.

(ii) Mr. Ross blames relativists for not having exhausted all the possibilities of the older theory. On their own admission all that we directly know is that the earth and the stars move with respect to each other. If there be an ether this fact is quite compatible with the earth being at rest with respect to it. Now the results of the Michelson-Morley experiment are paradoxical only because the earth is assumed to move through the ether, not because it moves with respect to the stars. And the latter, we have seen, does not imply the former. Mr. Ross's alternative would split into two forms according as he holds: (a) that there is, or (b) that there is not relative motion between different parts of the ether. On the former alternative both the earth and the stars might be at rest relatively to the parts of the ether in their immediate neighbourhoods. On the latter alternative the stars would have to be moving through the ether and to have the same velocity with respect to it as with respect to the earth. The former hypothesis has been tried, and is known to lead to conflicts with the facts about aberration. The latter, I think, is the one that Mr. Ross has in mind. It cannot be regarded as plausible to hold that the earth is the one body at rest in an ocean of stagnant ether, whilst the stars are all moving about in it. If the ether be a real physical substance pervading the whole universe, as those who take it seriously enough to entertain either of these alternatives must hold, this second alternative places our small planet in a strangely unique position. But apart from these *a priori* objections, the physical difficulties in any such view are colossal. To account for aberration we shall have to suppose that all the stars describe ellipses in the ether in the period of a year. These ellipses will have to be adjusted to each other in a very intimate way, for which the present theory supplies no explanation. Moreover, considering the extreme remoteness of many of the stars, the ellipses will be of gigantic size, and therefore the velocities with which the stars must move in order to describe them in a year will be stupendous in some cases of the same order as that of light. Not only are the dynamical difficulties of supposing such large masses to be in such swift motion very great, but the shifting of the lines of the spectrum in light from such stars, due to the Doppler effect, would, I imagine, make stellar spectra utterly different from what they are found to be.

(iii) But Mr. Ross's main difficulty is that he thinks that relativists take absolute motion as a premise in their proofs of the relativity transformations, and that these results are then supposed by them to disprove absolute motion. Before considering in detail whether relativists actually do this we may point out what exactly would be the logical consequences of such procedure. If the observable facts and the assumption of absolute motion imply the relativity transformations, and these in turn imply the denial of absolute motion, it will follow that the facts and the assumption of absolute motion imply the denial of absolute motion. From this we *should* be justified in going on to deny absolute motion. But we should *not* be justified in taking the further step of asserting the theory of relativity. Thus, if the relativistic arguments were of the form which Mr. Ross believes, and if there were no internal fallacy in them, we should be justified in denying absolute motion but not in asserting the theory of relativity.

Actually, however, Mr. Ross is mistaken in thinking that relativists use the absolute theory as a premise to prove the theory of relativity. Let me take my own case, e.g., as Mr. Ross accuses me of this procedure. For didactic purposes I started with the ordinary assumptions of absolute space, time, and motion, and an ether at rest in this space. I then drew a distinction between distances, time-lapses, etc., and our *measures* of these. And I showed that if we wanted to account for such facts as the Michelson-Morley on these assumptions we should have to assume certain physical changes in our rods and clocks when they moved through the ether. The results of these changes are summed up in the transformation equations, and at this stage these may be regarded as expressing the connexion between the distances and time-lapses which we should record if our system were at rest in the ether and those which we should record if we were moving through the ether with an uniform rectilinear velocity. At that stage I was not attempting to prove the *theory of relativity*, but only to prove that such and such relations must hold between our readings when we are in motion and the absolute magnitudes if the facts are to be squared with the absolute theory. The next stage is to reflect on these results. (a) We see that the physical processes needed to make the absolute theory square with the facts are unnatural in the last degree, and that they have neither the causes nor the consequences which such processes might be expected to have. (b) We notice that, since the result of the transformations is that the measured velocity of light will be the same for all systems in uniform rectilinear motion, we may just as well interpret the c of our formulae as that relative velocity and drop all reference to the velocity of light with respect to the ether, which was its original meaning. (c) Next we notice that the form of the equations is such that the transformations from one system to another in uniform relative motion will be precisely the same as the transformations from a given system in motion to one at rest in the ether. We have merely to substitute everywhere in the formulae the velocity of one system with respect to another for the velocity of a given system with respect to the ether. We can thus reinterpret the v of our formulae provided we make a parallel reinterpretation of the x, y, z, and t. The v is now to stand for the velocity of one system as judged from a second, instead of the velocity of a single system with respect to the ether. The x, y, z, t are now to stand for the measures of length and time-lapses found by people on the second system, and the transformation equations give us the corresponding measures of length and time-lapse found by people on the first system. Thus absolute motion and the ether have dropped out altogether, and we are left with equations connecting the measurements of two observers who contemplate the same events. Had absolute motion been a *premiss* for proving these equations, of course we should have no right to reject the premiss and hold that we had proved the equations. But the real position is that the evidence for the equations is simply and solely that they account for the facts. If there be absolute motion it must have such physical effects as to lead to these relations between the measures found by two observers in uniform relative motion, for these relations are found to be necessary to explain the facts. But on the one hand, if there be no such thing the relations will still hold. And, on the other, the facts that absolute motion in any case cannot be observed, that it cannot be inferred from its effects because these are such as never to show themselves, and that the effects which we should have to ascribe to it accord very ill with the rest of our knowledge of nature, strongly encourage us to try to dispense with it altogether.

(iv) The last point in Mr. Ross's paper on which I want to comment is his remarks on simultaneity. His view is that we all know what simultaneity means, and that it always means the same thing. Einstein gives a *test* for it in certain difficult cases, this is never a *definition*, and as such it may be right or wrong, while a definition could only be convenient or inconvenient. I agree in part with Mr. Ross here; but I do not think that the point at issue is so important as he makes out. Certainly I do not primarily *mean* by simultaneity anything to do with light signals. And I do mean *something* by it. But (a) I may mean something by a word and not know all that I mean by it. I may think it stands for an absolute term whilst it really stands for a relative one. I talk, for instance, of the colour of a piece of gold and only learn afterwards that the colour is not a property of the gold by itself, but is relative to the physical situation in which the gold is placed. Similarly the fact that I mean something by simultaneity, and think that it is an absolute term, is quite compatible with its really being relative to a co-ordinate system. I think the colour of gold to be non-relational because I tacitly assume certain familiar conditions of illumination which are normally fulfilled. In the same way I may fail to notice that simultaneity has an essential reference to a co-ordinate system because I habitually assume a certain familiar system. It does not seem to me that we start life with a clear enough knowledge of what precisely we do mean by simultaneity to deny this off-hand. (b) Granted that we may mean something by a word without knowing with perfect definiteness what we do mean by it, and that this uncertainty allows the possibility of its standing for a relational term, I think Einstein is justified in assigning any meaning to it in doubtful cases which does not fall outside the range of variation of our meaning. He then naturally choses that particular meaning within this range which allows of a definite test and simplifies the statement of the laws of motion as much as possible. This is a general procedure in all sciences, and seems to me to be a perfectly legitimate one. We are not, as Mr. Ross thinks, claiming to give a perfectly arbitrary meaning to a previously meaningless noise; the noise has a restricted class of possible meanings, and we are choosing the most convenient and reasonable one within this range. (c) Lastly, if it be granted that relativity *to a co-ordinate system* falls within the range of possible meanings of simultaneity it follows that such relativity as is found need not be to *our minds* or our judgments, as Mr. Ross seems to think. And the fact that we are not dealing here with a relativity that merely refers to our minds and their judgments is proved by the fact that purely physical systems, such as spectroscopes or the moving liquid in Fresnel's experiment, themselves 're- cognise' the relativity transformations.

I hold then that, even when we were confined to the special theory, we had good grounds for viewing it with great favour, and that we committed none of the fallacies of which Mr. Ross accuses us in our arguments for it. But I think the general theory is in an even stronger position than the special theory. Let me explain just what I mean by this. Mr. Ross says he will confine himself to the special theory, because, until one has convinced oneself of it, it is useless to worry about the more general one. This seems a reasonable attitude to take, and yet I believe that it unconsciously does an injustice to the theory of relativity. The general theory has in its favour all the arguments that favour the special one, and in addition, certain arguments which do not apply directly to the latter. These arguments consist in the extraordinary unification which it introduces into physics, and the way in which it removes that deplorable scandal which had always hung over the Newtonian laws of motion. The unification of course is that it binds together in a single whole Newton's two great achievements, the laws of motion and the law of gravitation, and connects the two previously independent notions of gravitational and inertial mass. The scandal was the necessity of a particular frame of reference for Newton's laws. If you took this to be absolute space you had laws which were presumably discovered by observation, and intended for application to the empirical world; and yet they were stated in terms of entities which could neither be observed nor inferred. If you took the frame to be the fixed stars you felt that they were placed in an utterly unintelligible position of importance in nature. It seemed obvious that there must be some way of stating the laws of nature on the one hand entirely in terms of relative motions and positions, and on the other independently of some one special group of material objects such as the fixed stars. To have done this is the great service of the general theory and the overwhelming argument in its favour, to my mind.

To sum up as regards the evidence for the theory: It seems to me that the general theory starts by shocking us through its unfamiliarity, but that the more we reflect on it and on the mass of perfectly gratuitous and essentially unverifiable assumptions involved in all the alternatives the more certain do we become that it, or something extremely like it, must be true. If men like Prof. Eddington or Prof. Lindemann, who have been constantly and successfully using the methods and results of the theory, were the only people to make the above statement, we might be inclined to discount it somewhat as expressing 'the bias of happy exercise.' But the fact that I am a mere philosopher, quite incapable of their mathematical and physical achievements, may at least serve to allay such suspicions when the statement comes from me.

I will conclude with some remarks on Prof. Eddington's most interesting theory as to the function of the mind in physics. I will not call them criticisms, but rather appeals to Prof. Eddington to clear up some places where his meaning seems to be doubtful. (i) He often speaks as if lengths, time-lapses, etc., were relations between Nature and the observer. He thus seems to make Nature simply the almost unknown referent of these and other relations. Would it not be nearer the truth to draw a much sharper distinction between the 'observer' in the sense of his body and his scientific instruments and the 'observer' in the sense of the observing mind? In the former sense the observer is part of nature, in the latter he is not. And we ought then to say that lengths, time-lapses, etc., are relations between one part of nature and another part of nature, and it is these relations – or the natural complexes related by them – which the mind of the physicist contemplates, measures, and describes, (ii)

I am not sure that Prof. Eddington does not state his selection theory in needlessly subjective terms. To take a crude illustration: Suppose that a number of dots were scattered about at random on a plane. Any three of them would form a triangle and any four of them would constitute a tetragon. The triangles and the tetragons are equally real, and equally parts of nature, and you could completely analyse nature into either. But, on the other hand, only a small number of the points, if any, might be at the corners of squares. Now let as suppose that both triangles and tetragons have properties corresponding to 'conservation.' Then the whole of nature could be analysed exhaustively into entities obeying laws of conservation. If, on the other hand, only squares had the property corresponding to conservation, then, however much the mind might be interested in conservation, it could not give an exhaustive account of nature in terms of conservative entities, and it might be the case that *nothing* in nature obeyed such laws. Now the question I want to ask Prof. Eddington is this. Can *any* four-dimensional manifold be exhaustively analysed into complexes having the property of conservation, as *any* set of points in a plane can be exhaustively analysed into triangles or tetragons? If so, of course, the fact that nature everywhere obeys laws of conservation is in no way due to the mind but to the properties of four-dimensional manifolds as such. The result would be that such laws are necessary in all possible four-dimensional worlds. If not, then the important question would be: Does the actual four-dimensional world in which we live admit of exhaustive analysis into subordinate complexes of this special kind? The fact that the mind happens to like such complexes would of course throw no light on this question. The fact, if it be a fact, that it neglects all other complexes and yet seems able to describe and deal with nature satisfactorily would suggest that probably this condition is pretty nearly fulfilled. For, if there be other complexes and we be so constituted that we neglect them, it does not follow that they will neglect us. And we should therefore expect to get into serious practical and theoretical difficulties if the bent of our mind caused us to ignore types of complex which are real parts of nature and cannot be analysed into the complexes of the types that we do notice.

Scientists generally and rightly neglect the existence of minds while going about their lawful business. When at a later stage minds are forced on their attention they tend to be embarrassed. If they be stupid they deny minds altogether, which seems to be the last asylum of the dogmatic biologist. If, like Prof. Eddington, they have too much sense to do this, they are liable to go to the other extreme and, taking *omne ignotum pro magnifico*, to ascribe to minds powers and functions which they probably do not possess. I do not assert that Prof. Eddington has made this mistake but I have my suspicions.

IV. By F. A. Lindemann

The difficulties of Mr. Ross seem to have been dealt with very completely by Mr. Broad so that I will confine myself to an attempt to restate the general case for Relativity in its simplest form in the hopes of providing a basis for discussion.

For this purpose I propose to examine the question why we study physics and attempt to establish the relation between physics and metaphysics. Then to state the impasse which led to the special theory of relativity, and finally to explain the essential difference between the general theory of relativity and the Newtonian point of view.

Mankind has evolved in the course of ages amidst hostile surroundings from the position of one of the minor fauna to that of unquestioned master. Whatever may be the reason for this we cannot therefore be surprised if man has many attributes of considerable survival value. There can be little doubt that one of the most valuable characteristics from the survival standpoint would be the faculty of forseeing future events, and it is not to be wondered at therefore that those races and men who have survived have an innate tendency, possibly strengthened by tradition, to seek to correlate events and establish relations between phenomena, which will enable them to predict subsequent happenings from observed data. The more easily such relations or laws are assimilated and applied, the simpler they will appear, hence the human mind, being what it is, always tends to accept the simplest laws consistent with observed facts.

Physical laws, and probably all laws, are based on observed phenomena. In order to establish a law a physicist observes a phenomenon under various conditions, formulates a hypo- thesis to account for the results, extrapolates new consequences of his hypothesis, tests these empirically, if necessary modifies his hypothesis, and so on. In this way, by a series of successive approximations he arrives at a rule or law or formula which is valid for all his experiments, which should be valid for all experiments carried out under conditions intermediate between those actually tried, and which is often valid when extrapolated for a considerable distance beyond the observed instances. A man with this physical habit of mind may occasionally be misled by insufficient data, but when this happens his constant empirical checks inevitably show him his error and cause him to recast his theory.

We may contrast with the physical habit of mind, which we all have to a greater or less extent, what may perhaps, for want of a better term, be called the strictly logical habit of mind which occasionally survives in universities and other secluded regions. A logician of this type refuses, at any rate in theory, to believe that it is possible to learn by experience or extrapolate from observed repetitions. In his view the fact that the sun has risen a million times in succession does not provide any reason for believing it will rise again. He says one must either know or admit ignorance, and deplores our tendency to simple extrapolation. One can imagine occasions upon which the logician might score at the expense of tne physicist who frankly admits that he does not know, but finds it pays to extrapolate, e.g., at Monte Carlo, where the logician should never even be tempted to invent a system; but in the infinitely more numerous and important affairs of daily life the physicist would survive whilst the logician would perish. Still some individuals with a tendency to this type of logic, or better still some chromosomes or chromidiae, which predispose an individual to such a dangerous habit of mind, have managed to survive. They have done this by making a new and perfectly undemonstrable assumption, namely, that certain things or laws are "self-evident."

Making such an additional assumption of course complicates things and thereby diminishes the probability of the survival of the individual characteristic; but it need not diminish it very materially if the "self-evident" truths are judiciously selected. Clearly any member of the congress who inclines to the "logical" point of view has survived and he would not be amongst us to-day unless his self-evident laws approximated to our physical laws. The danger of course lies in the fact that a "self-evident" law, once it rises above the level of a mere definition or tautology, is always liable to be upset by new experimental evidence.

Now the law may be "upset" in two ways, so different quantitatively that they may almost be considered qualitatively different, and it is this difference which, in my opinion, forms the only distinction between physical and metaphysical statements. As an instance of a physical statement, than which few things could seem more "self-evident," we may instance the claim "that water is continuous and homogeneous."

This involves the claim that it would be possible in principle to subdivide a drop of water into an infinite number of particles, each of which would have the properties of water. We have every ground for believing that if it were possible to cut a drop into eight equal parts by three perpendicular cuts, and repeat this process some twenty-five times, we should arrive at something very different to water, namely, hydrogen and oxygen. To refuse to believe this because the continuity of water appears self-evident would practically amount to repudiating the whole edifice of modern chemistry and physics. The number and complication of supplementary hypotheses that would have to be made in order to take account of observed facts, would be so enormous that a physicist must refuse to contemplate such an alternative.

As an instance of a self-evident truth of the second type we may take the geocentric system of cosmogony "self-evident" if anything can be. Why was this system superseded by the heliocentric system against the tradition of centuries, the authority of religion, and the efforts of the secular powers? Only because the Copernican system is simpler. Both systems are capable of accounting for all the facts, and it is really surprising how quickly the simpler theory supplanted the more complex merely by virtue of its simplicity against all the weight of prejudice, and in spite of its "self-evident absurdity." Its acceptance is an inspiring proof of the innate tendency of the human mind to assume that which is simple and manageable, and which therefore tends to the preservation of the race.

The difference between the two examples is clearly one of degree rather than of type, but the difference of degree is enormous. The geocentric system could be worked, though with more effort than the heliocentric. The denial of the discrete nature of matter would probably involve complications which would transcend the capabilities of the human mind. From this point of view, therefore, a physical statement is one which it is impossible to give up without revolutioning science, whereas a metaphysical statement is one which forms a convenient basis for describing phenomena, but which has scarcely more importance intrinsically than has the choice of co-ordinates in geometry.

It is difficult, if not impossible, to say just how much gain in simplicity is necessary in order to justify us in believing that a certain theory is intrinsically true rather than merely convenient. Here again we must trust to the inherited tendency of the mind to draw the line. But most people will agree that there is a vast difference between assuming, say, that the earth is round, because this is the simplest way of accounting for the observed facts, and assuming that the earth is divided up into parts by lines of latitude and longitude because these provide the easiest way of specifying a point on the earth's surface.

In my opinion the Principle of Relativity is what has been defined above as a metaphysical principle, and we are now in a similar position in respect to the theory of Einstein that Galileo occupied with regard to the cosmogony of Copernicus. We find it hard to give up our prejudices in favour of a strict distinction between space co-ordinates and time co-ordinates, and in favour of a strictly Euclidean space merely because it simplifies the laws of physics. To do so requires a mental effort which, in the opinion of some, is not compensated by the gain in simplicity which results. But our notions of space and time are essentially metaphysical conceptions, and as such are clearly merely a matter of convenience or even of taste. The older generation may, therefore, be justified in refusing to accept the new doctrine and sticking to its "self-evident" truths at the expense of simplicity, but as in the astronomical parallel we must look for progress and discovery to those whose elasticity of mind enables them to make themselves familiar with the new point of view. Neither standpoint can be said to be right or wrong since cither enables us to represent the facts adequately, in fact, as mentioned above, the difference is not so very much greater than one of a choice which co-ordinates one will adopt. But the old theory panders to outworn prejudices at the expense of simplicity, whilst the new will probably seem as obvious and natural in a generation as the Copernican theory does to us to-day. Just as the change from the geocentric to the heliocentric cosmogony denoted a momentous emancipation of the human intellect, so does a grasp of the theory of relativity enable us to look with a much wider and broader view on the systems and philosophies of the past.

As a basis for discussion it may be worth while to set down once again in the baldest form the experimental facts which seem to show the desirability of reconsidering our opinions, firstly, as to the sharp distinction between space and time co-ordinates (special theory of relativity) and secondly, as to whether space, or if the first thesis be accepted the space-time manifold, is Euclidean (generalised theory of relativity).

Perhaps a brief, almost historical, analogy may be interposed, which illustrates the situation which led up to the special theory of relativity. Let us picture a primitive community in which height is rigorously distinguished from length and breadth. This distinction might well appear fundamental since work must be done in order to raise an object, whereas it can be moved in a horizontal plane without effort. As long as the members of the community believed the earth to be flat, they would consider it just as easy to distinguish height from the horizontal dimensions as we tend to think it is to distinguish time from the spatial dimensions.

Now suppose an observer on the top of a tower observed a distant tower with a theodolite. If both towers were of equal height when measured in the usual way by means of a plumb line and a foot rule, our observer would expect to find that his theodolite was level. On account of what we call the depression of the horizon he would of course find that he was obliged to point slightly downwards. At first he might attribute this to some peculiarity of the air, but when he found the same phenomenon whichever eminence he ascended, he would be forced to seek a more general explanation. The first that would occur to him would probably correspond to the Lorentz-Fitzgerald contraction. He might say that the mere fact of ascending distorts the scale of the theodolite and elaborate a consistent but complicated system on these lines. A really clear thinker, who would free himself from prejudice, might however proceed as follows. He would say, this distant tower is lower than mine for my theodolite measures its height. Therefore when I drop a plumb-line from it and measure the length the plumb-line cannot be parallel to my plumb-line in my observatory. But my observatory is in no way pre-eminent above any other spot in the world, therefore I cannot say my plumb-line is truly vertical and measures height, whereas all others are deflected towards me. Hence the direction which we call height must vary according to which part of the earth's surface we are at and what I call height must appear to be composed of height and horizontal distance, for anybody else and *vice versa*. The simplest way in which I can express this is to say that the surface, which I have been taught to call plane is curved and to say that height is the direction normal to this surface.

It is not necessary to picture the scepticism with which such an argument would be met in detail, how the unfortunate originator of the theory would be told that everybody knew what height, was and that to try and compound height and horizontal distance was as foolish as to mix space and time, and how he would be finally overwhelmed by some philosopher pointing out that his theory logically involved the possibility of Antipodeans. Such a description would apply to events even yet too recent to be altogether pleasant. But though the analogy is obviously imperfect the results of the Michelson-Morley experiment put us in a very similar predicament to that pictured above.

Unless we assume that the earth is altogether pre-eminent in the universe and that the Michelson-Morley experiment, which yielded a purely negative result on the earth would show a positive result on any and every other planet, we can describe it, making use of Majoranas' results, in the following way.

If two observers moving past one another sent out a light signal at the moment they are in contact this signal will spread out as a shell of light. Although they are moving away from one another, each observer will find as the result of the most accurate measurements that he is and continues to be at the centre of the expanding shell of light.

If the shell of light has objective reality there is only one explanation for this, namely, that the standards of length and time used by the two observers, which agree when they are at rest relatively to one another, do not agree when they are not at rest relatively to one another. If the two observers A and B are moving with the relative velocity V it is easy to specify the exact change of the units of length and time which would lead them both to conclude, as really happens, that they are at the centre of a spherical shell of light expanding at velocity C. This change is expressed uniquely by the Lorentz transformations and is such that A considers that B's measurements of length involve what he, A, calls length and time, whilst B considers that A's measurements of length involve what he, B, calls length and time. The same holds good for measurement of time. Each observer finds that the other observer must be measuring a quantity involving both time and length when he thinks he is measuring time.

Now no observer is pre-eminent above any other and therefore neither can claim that he is right and that the other is wrong. Each considers he is separating length and time in the one obvious unique way and yet neither is separating them from the other's point of view. The obvious conclusion is that they are both viewing the same event in a four-dimensional space-time manifold from a slightly different angle. This is precisely what the equations of transformation which may be found as shown above indicate to the mathematicians.

An event implies both spatial and time relations and in order to describe it we introduce space and time co-ordinates and represent it in a four-dimensional manifold. The achievement of the special theory of relativity consists in having shown that there is no unique way of separating space and time co-ordinates but that observers moving relatively to one another will separate them in different ways. Objective reality belongs to the event, its description in terms of space and time varies and depends upon the observer: space and time are thus relegated to the secondary role of convenient co-ordinates personal to the observer which he uses in order to describe events.

The main philosophic advance to be claimed for the generalised theory is to the emphasis it has laid upon the fact that the conceptions we choose to form about geometry in the four-dimensional space-time manifold which forms our universe are entirely arbitrary. Again it is purely a matter of convenience which geometry we adopt.

There is no meaning in saying any particular geometry is true or false, that space is Euclidean or non-Euclidean, homaloidal or not, for space without objects is inconceivable. Therefore any statement about space really consists in a statement about objects, preferably solid objects. It is readily seen that here a wide range is open. Thus if anybody chooses to affirm, for instance, that the linear dimensions of all objects in a room contract to one-half when turned from a N.S. direction to an E.W. direction, it is impossible to prove him to be wrong. Clearly his measuring rod will con- tract by the same amount so that the fact that the measured length does not alter proves nothing. The only objection to such a scheme is that it involves complicated laws of physics.

Take, for instance, the elementary mathematical treatment of a game of billiards on these assumptions. Two balls moving E.W. and W.E. may be made to collide at such an angle that their directions are changed to N.S. and S.N. respectively. Neglecting friction their speeds appear to remain the same. But if we assume that the E.W. dimensions are one-half of the N.S. dimensions the speeds, which appear unchanged, must really be doubled and the kinetic energy must have increased to four times its original amount. If we desire to make the above assumption about our geometry, or space, or perhaps best of all about the properties of solids, and yet retain the laws of the conservation of energy and momentum we can only do so by making special assumptions, e.g., that E.W. kinetic energy is four times as great as N.S. kinetic energy and E.W. momentum twice as great as N.S. momentum. Similar arbitrary assumptions would be required in order to account for other phenomena, but there is no doubt that a consistent system of laws could be evolved to fit an anisotropic space. The objection is, of course, that such a system would be very much more complicated than the system we use. In view of our innate tendency to adopt the easiest and simplest system therefore we usually assume space to be homaloidal.

For the same reasons we have hitherto assumed space to be Euclidean, namely, because this appeared to lead to simple and convenient laws of physics. It was Einstein who first pointed out that even simpler laws result if we give up this assumption which long usage has rendered almost a necessity of thought to some minds. The simplification is perhaps best seen if one tabulates the postulates necessary to account for observed facts in gravitational physics on the bases of Newton and of Einstein.

From the absolutist point of view we must assume:

1. That bodies unaffected by other bodies follow the shortest paths, i.e., that their four-dimensional world lines are unique.

2. That space is everywhere Euclidean.

3. That bodies or energies attract one another with a force proportional to the product of their masses.

4. That this force varies inversely as the square of the distance.

5. That a quasi-magnetic force acts upon bodies or energies moving relatively to one another.

From the relativist point of view we must assume :

1. That the four-dimensional world lines of bodies or energies are unique.

2. That the curvature of space is proportional to the mass.

3. That it is inversely proportional to the distance.

The absolutist system introduces a mysterious entity called force and requires five assumptions at least. The relativist system yields all the same results with but three assumptions. The latter, therefore, appears preferable, but to say that one assumption is true and the other false would be just as meaningless as to say that space is or is not homaloidal. Either point of view is perfectly justified, but the one appears simpler, and, therefore, more convenient than the other. It would be unwise, though nobody could say it was wrong, to attempt to use Cartesian rather than polar co-ordinates in discussing curves such as spirals. If a mathematician existed who had never studied trigonometry or heard of polar co-ordinates, he might consider it better to treat the problem in this way, in spite of the complication, rather than make the mental effort necessary in order to familiarise himself with a new world of sines and radii vectors.

No man can estimate his neighbour's mental elasticity, and no man, therefore, has the right to condem another who refuses to embark upon an adventure for which he must dispense with the sword of self -evidence and the armour of prejudice, in which most of us are entrusted, and rely upon forging new weapons as he goes along. Each man must be the judge of his own limitations. But there seems little doubt that the future will belong to those who are able to realise when their mental accoutrement has become so unyielding as to be more of a hindrance than a help, and who have the courage and initiative to cast it aside and adopt new methods rather than wait until their own have been superseded.

www.ingramcontent.com/pod-product-compliance
Lightning Source LLC
Chambersburg PA
CBHW071213200326
41519CB00018B/5511